NORA DUNCAN O'BRIEN

Navigate Family Technology

A roadmap for family technology use with ideas to navigate uncharted waters

For Colin, Adrian, Dylan, and Julie Scarlata, whose patience on this journey has known no bounds, and I am forever grateful for their help.

"As a computer scientist, I make a living helping to advance the cutting edge of the digital world. Like many in my field, I'm enthralled by the possibilities of our techno-future. But I'm also convinced that we cannot unlock this potential until we put in the effort required to take control of our own digital lives—to confidently decide for ourselves what tools we want to use, for what reasons, and under what conditions. This isn't reactionary, it's common sense."

-Cal Newport

Contents

Foreword

My mom writing a book was definitely an interesting experience. There were lots of different challenges for her but also
for me and my brother. We were a little bit like the guinea
pigs of the research. At first I didn't like it and thought it was
unfair that I got less screen time than other kids and did not
get a phone until I was thirteen. I then read *Navigating Family
Technology* and it all made sense. At first I thought my mom
was making some of the stuff up, but then I realized that all
this information is what experts say to do. I learned a lot and I
hope parents and kids alike can too!

~Nora's son, age 14

Introduction

Modern tech is designed to draw you to your device like a moth to a flame. You were born to connect, and your devices know it. Apps and platforms offer an inviting promise of connection while subtly pulling you into your own world, fueling a rising tide of loneliness and disconnection across modern society.

For millions of years, moths, butterflies, and dragonflies relied on natural light for navigation; moonlight and starlight have guided them. The introduction of artificial light creates confusion: their most reliable instinct works against them by drawing them toward a brightness that offers no destination; they circle endlessly, exhausting themselves in pursuit of a light they weren't built to handle. The light isn't the enemy; their wiring is exploited by something their systems weren't designed to encounter.

Our human desire to connect has driven us for generations. Today, a glowing screen seeks to fulfill that need for connection with speed, convenience, and endless stimulation. It tells us, *this is what you're looking for.*

This wiring is powerful, and it is being expertly engaged by tech companies. Whether you're raising children in today's digital landscape or trying to navigate it yourself, this book is not about rejecting technology. It is about understanding it.

Navigate Family Technology will help you recognize how your instincts are being shaped, where your attention is being pulled, and how to reclaim intentional connection at home. You'll gain practical strategies, a grounded perspective, and the reassurance that you are not alone.

We're more reachable than any generation before us. And yet we're lonelier.

The parenting adventure has changed dramatically since today's parents were raised. For those of us without childhood experience in a high-tech world, technology can feel like an invisible wall between our kids and us. Parents of babies, toddlers, preschoolers, school-aged kids, tweens, and teens report the same tension: how do we balance our instinct to limit screen time with current societal norms and our wildly busy lives?

Modern tech has many upsides: the ability to stay in touch with people outside your inner circle is modern magic. Connecting with those who share similar interests fuels camaraderie. Technology is a gift that expands knowledge and offers clarity on difficult concepts. While acknowledging tech's upsides, awareness of and intention around its downsides are crucial to well-being.

I wish I could say I wrote this book because I had it all down -I'd figured out and perfected how to raise kids and live a perfect life in today's world. I wish I were not prey to modern tech's allure. I began exploring this topic when I realized the magnetism of tech was tightening its grip on my family and me. Subtopics were added as friends and acquaintances shared tech challenges surfacing in their lives.

Professor and best-selling author Jonathan Haidt is making a global impact with his work on kids and screen usage. Much-needed changes are taking shape: schools banning classroom device use, parents delaying child smartphone access, child-hood free-play efforts on the rise, and legislation to raise social media age limits underway in parts of the world. While Haidt tackles the overarching societal shifts, I write to help parents navigate the day-to-day challenges modern tech presents.

Reflection and choices around technology use begin when kids are babies and continue throughout our parenting journey. Elementary school and adolescent phone/device challenges do not occur overnight. The slippery slope of screen habits can soothe, pacify, distract, individualize, and eventually disrupt connection with the outside world. Technology utilized with awareness and intention optimizes outcomes.

Every child is unique, yet many families share comparable, unwelcome drama around screen use. Researchers have now gathered enough data to draw clear conclusions about kids and technology. The findings point to technology use as a significant factor in the rise of childhood and adult stress, anxiety, and depression.

Though some experts advocate a total ban on device use until adulthood, in my experience, that would contribute to societal disconnection. This book offers a more nuanced approach, focusing on continued thought about what healthy device use looks like as adult role models and how parents can choose to captain the ship of device usage within their families to positive outcomes. Families can live intentionally rather than in a constant reaction mode to the difficulties modern tech may bring.

Navigate Family Technology provides deep dives into tech top-ics, including tech's influence on focus, empathy, anxiety, and more -a roadmap for today's technology landscape. *Roadmaps are a resource, not a planned route,* akin to a choose-your-own-adventure.

The goal is not a perfectly manicured path, as there is no one-size-fits-all approach. The chapters can be a springboard for consideration of tech's role in our lives and the conversations we have about it.

We cannot slap a return label on modern tech. We can understand it, push back on it, use it to our benefit, and reclaim human connection. *Navigate Family Technology* arms families with information on modern tech's downsides and illuminates countless strategies to overcome them.

Chapter Guide

Knowledge for Parents with Kids 0-8:

Chapters 1, 2, 3, 5, 6, 11, 12, 13, 14, 15

Knowledge and Conversation Starters for Parents with Kids 8-14:

Chapters 1, 2, 3, 4, 5, 6, 8, 11, 12, 13, 14, 15

Knowledge and Conversation Starters for Parents with Kids 14-18:

All Chapters

Knowledge and Conversation Starters for Parents with Young Adults Ages 19-27:

Chapters 1, 2, 3, 5, 6, 8, 9, 10, 11, 12, 13, 15, 18

General Knowledge for Any Adult

All chapters

Parents of tweens, teens, and young adults have shared that reading some chapters together (or simultaneously) as a family has eased their burden on difficult conversations. Reading chapters together sparks rich discussion and can facilitate thoughts about family values.

Author's note: My favorite chapters are chapters 10-15. Many have mentioned that Chapter 8 was valuable to them. Chapter 18 has been fully updated and might become the most important chapter as we venture forward into an AI-filled world. Enjoy in whatever order you wish, by age, interest in topic, or whatever you're feeling -Nora

1

Chapter 1: Family Communication

"Technology gives us power, but it does not and cannot tell us how to use that power. Thanks to technology, we can instantly communicate across the world, but it still doesn't help us know what to say." -Jonathon Sacks

Parental challenges have changed. Technology has become a source of friction between parents from an era before iPads and smartphones, and kids who were born into an advanced tech world that changes every day. Parents are flying blind, learning as they go, trying to balance letting kids be somewhat in line with their peers and learn from their mistakes while protecting and guiding them as they navigate the inherent problems and dangers that come with modern technology.

How and what we communicate make a difference. What if we don't have to reinvent the wheel and live in constant reaction mode?

Whether younger kids are on tablets watching endless

YouTube videos or older kids are drawn to phones buzzing with constant messages, social media posts, or online gaming, it is clear that technology rules our kids' lives. Family battles over technology seem unavoidable, so how can parents stand their ground without seeming unfair, old-fashioned, or out of touch?

Changing Roles

Parent Role Changes

01   Little Kids

Parents are intern supervisors/ new hire trainers. Heavy oversight, explanations, guidance.

02   Middle Childhood

Parents are managers - kids make some decisions, begin to have more autonomy and more of a say. They learn problem-solving skills and grow from failures and experience.

03   Adolescent and Early Adulthood

Parents shift into roles with less oversight and supervision, more like a CEO or consultant.Parents give kids more autonomy to make decisions, and there is a support system in place to oversee the big picture and assist.

Kids and parents morph through role changes as kids grow

Many factors will influence kids' decisions from here on out, including life experience and their peers. CEO's hope that seasoned employees have absorbed core messages and key company values.

For me, intention about technology use is a core value I hope sticks. If that's the case for you, conversations that begin in the early years and continue throughout emerging adulthood can impact kids' current and future choices. Gaining kids' buy-in on intentional tech-use will be key.

We've confused access to people with connection to people. Passing that message to younger generations has the potential to reverse the steep increase in loneliness and disconnection.

Chapters that speak to you might be sparks for family communication when appropriate.

Communication Style

Apps and devices change at dizzying speed, delivering an exhausting game of whack-a-mole for modern parents. Staying on top of every latest development is impossible. Clear, open communication channels are what help today's families navigate these times.

Regular, ongoing discussions are key. Families can sprinkle in conversations about screen use and stay ahead of issues that arise in homes worldwide. Open, proactive communication can set families up for better outcomes on common challenges. Short, balanced conversations diffuse negativity, emotion, and the friction these topics bring. Particularly, for more difficult conversations as kids get older, brief, *ongoing* conversations work best.

Shoulder-to-shoulder communication works well, especially with tween or teen boys on sensitive topics, advises Digital Wellness Coach Katey MacPherson. MacPherson says, "I would put them in the car and make up an errand. Especially if it's a boy, shoulder-to-shoulder communication is going to give you so much more mileage in that talk than trying to talk face-to-face. The [crosstalk] actually lights up the male brain with cortisol and shuts down the verbal emotive centers. Especially with our boys, it's super important that we're not doing this and asking them four hundred questions. They are not going to give you much."

Families might initiate discussions about screen use throughout society and the world and bring it back to their families. Asking for kids' thoughts on tech topics creates conversations rather than lectures.

10

Outings to observe interpersonal communication at restaurants, on walks, in movie theaters, and more can spark conversations about how others may feel when their friend or peer pulls out their phone during in-person connection time.

Positives First

Talking about tech positives before tech negatives can create a team feeling around the subjects rather than the battlefield tech discussions often bring.

Parents can learn about kids' tech with them -discover what they love about it! Younger kids relish the opportunity to walk parents through their favorite videos or games, showing off funny or interesting items they found, won, or built. Teens may talk about online personalities they follow.

Validate what kids love about tech before sharing tech concerns. What do you love about modern tech? I love Shazam, my inexpensive fitness tracker, access to a camera, and birthday text messages. I greatly appreciate a digital weather update before heading outside.

Relationship gurus John and Julie Gottman remind their audience that relationships thrive on a five-to-one positive-to-negative communication ratio. When loved ones hear more negative than positive, they tune out and turn away. Keeping this ratio in mind during conversations about screen time can be the difference between a lecture your child endures and a dialogue that sticks.

Self-Determination Theory

Richard Ryan and Edward Deci's acclaimed Self-Determination Theory proposes that mental well-being is based on three things: **autonomy** (feeling in control of decisions and behavior), **competence** (continual growth through mastery of attainable but challenging tasks), and **relatedness** (feeling connected to others, a sense of belonging and inclusion within a group) by decreasing in-person connection.

Modern tech significantly lowers relatedness, driving stress. If tech rules are forced upon kids without an understanding of the why behind them, it diminishes their autonomy. Overuse of devices interferes with time for competence activities. Devices aren't inherently bad, but guidance on usage is necessary.

Continual family communication about technology can help kids gain an understanding of its pros and cons, which will ideally enhance their buy-in of why today's kids and adults have to work with intention to take control of their digital lives.

Communication by Age

Young children often look to parents as their trusted source -enjoy this communication phase!

Ages nine to eleven are prime times to relay important messages to kids -they are mature enough to understand more complex information, and though they begin to question parental judgment, they are not fully into the teen phase, which often involves pushing back on or doubting parental ideas.

For **parents of teens** (or tweens emitting teen vibes), certain conversation styles may help:

"I know you're a smart kid."

"What do you know about...?"

"What do you think you should do?"

"You've probably got some good reasons why you want to_____. Tell me about them."

"I've heard _____. What ideas do you have about it?"

"What are people saying about____?"

"Tell me more."

"I get how it's frustrating when other families are making different technology use choices."

Delivery

Adding playfulness and humor to conversations with kids of all ages can tone down anxiety and discomfort.

Tweens and teens are wary of feeling manipulated. Ironically, the same wariness that makes teens resist parental advice can become an asset -sharing how apps and industries are manipulating *them* often gets kids' attention fast.

Teens often prefer information from a trusted family friend, aunt, uncle, coach, or teacher rather than from their parents. It takes a village.

Remember the CEO metaphor? Even the best leaders welcome feedback, listen to their team, and give people room to grow, but they still set the guidelines. The same applies at home. Kids' input is welcome, encouraged, and heard. And like seasoned employees, as they get older and earn trust, they get a bigger seat at the table.

Parents may fear admitting they don't have all the answers, thinking it will make them look weak. In fact, this acknowledgment often gains trust and strengthens communication and buy-in. Families will have to continually adapt and change guidelines in the rapidly evolving modern tech world.

Changing the Playbook

One of the most common questions I get from parents is: how do we change course when the rules have already been set, or when there were no rules at all?

No coach would keep running a playbook from twenty years ago when the game has fundamentally changed. No navigator would insist the Earth is flat once better information proves otherwise. We update our understanding, and we adjust. Parenting in the digital age is no different.

Many of us gave kids devices without fully understanding the implications. We weren't negligent -we were doing the best we could with the information we had. We finally have ample research, the data is in, and we now know more about how these tools affect developing brains. When we learn better, we can do better.

Kids may push back. They may call it unfair. They may point out that the rules used to be different. That winds up as a great opening for an honest conversation: "You're right -the rules were different. Here's what we've learned since then, and here's why we're making changes."

Frame it as something the whole family is doing together, not something being done to them. New information calls for a new playbook, and everyone on the team has a role. Some ideas for introducing the shift:

NEW
PLAYBOOK

1 Transparancy

Share what you've learned and the why behind possible changes. Kids respond better to "here's what the research shows" than "because I said so."

2 Acknowledgement

The old rules weren't wrong at the time; they were based on what you knew then.

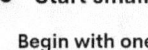

3 Invite

Kids into the process. Ask them what they think reasonable limits look like. They're more likely to buy into rules they helped shape.

4 Expect Resistance

Stay the course. Change is uncomfortable. Discomfort is not a sign you're doing it wrong.

5 Start small

Begin with one or two changes -not overhauling everything overnight. Small, consistent shifts build momentum without triggering rebelllon

Remind kids (and yourself) that every family is figuring this out in real time. There is no playbook from a previous generation to follow. You're writing it together.

15

The best coaches don't apologize for updating the strategy. They explain why the adjustment gives the team a better chance to win.

2

Chapter 2: The Magnetic Pull of Tech

"Do you know how many times an hour he checks Instagram? He has his phone in his hand and pulls it down to refresh the way a person might unconsciously flick a lighter, and then he glances down to see the new posts, even if he's in the middle of a sentence! He seems to have no idea he's doing it or that it's rude. You just can't take a person like that seriously."
-Unnamed Main Character, All Fours, Miranda July

Kids are addicted to technology. Adults are addicted to technology. Parents often complain tech usage is the biggest source of friction within families. Like tobacco companies before them, tech companies spend millions intentionally creating addictions; they hire behavioral psychologists to engineer devices and apps to win the battles for human attention.

Steve Jobs proclaimed that every kid should have access to iPads, then admitted he didn't allow his own kids to use them, maintaining high restrictions on tech in his household. Jobs' interviewer, Nick Bilton, writes, "Since then, I've met (many) technology chief executives and venture capitalists who say similar things: they strictly limit their children's screen time, often banning all gadgets on school nights." Many tech industry executives send their kids to device-free Waldorf Schools.

Netflix CEO Reed Hastings unabashedly admits that sleep is their biggest competition and brags his company is winning. Tech companies want eyes on the screen at the expense of mental and physical health activities, including quality leisure time, social bonding hours, productivity, and sleep.

How did tech companies succeed in making screens so addictive?

Three Drivers of Irresistible Screens

Adam Alter's three drivers of irresistible screen behavior:

1. **Endless play:** There are no cues to stop. No "Game Over" screen, ads, end of a show, or end of an article -just rebirths or extra lives in games, continuing to the next episode, for streaming, suggested for you on YouTube, bottomless feeds, and endless scrolling on social media. These features were designed after 2010, in an ethically questionable race to see which apps and media can hook

us most deeply. Today's kids are caught in the middle.

2. **Positive feedback:** We love rewards! We love accolades! We love presents! Likes, comments, re-posts, re-tweets, shares, loot boxes, free items in games, extra lives, etc, overstimulate dopamine levels and whet the appetite for more. Keep making us feel good, and we will stick with you.

3. **Goals:** Human nature finds satisfaction in progress and accomplishment. Tech companies make it easy to forget about real-life goals by substituting them for tech goals. They offer levels to achieve and streaks to keep alive -Snapchat communication streaks, Pokémon catching streaks, and opening an app streak. They challenge people to "win" numbers of friends, likes, and followers to watch and grow. Potentially diminishing real life productivity, goals keep kids (and adults) happy in their tech bubble.

"There's enough data that suggests we should be cautious about using screens, and especially cautious in how much time kids spend on screens," says Alter.

Vegas-Inspired Behavioral Psychology

Apps' "refresh screen" pull-down methods were modeled after slot machines in Las Vegas, the ultimate behavioral addiction. In her book, Addictions by Design, Natasha Schüll writes, "Facebook, Twitter, and other companies use methods similar to the gambling industry to keep users on their sites...In the online economy, revenue is a function of continuous consumer attention – which is measured in clicks and time spent."

Ease and regular access lay the groundwork for addictive behaviors. Residents within ten miles of a casino are 90% more likely to have a gambling problem than those who live farther away. We live awfully close to the devices in our pockets.

Author Michael Easter researches the brain -why humans crave things, and how profit seekers exploit these cravings. Easter learned about a "Human Behavior" research hub in Las Vegas that has unlocked the key to crushing human efforts at moderation and impulse control.

Easter elaborates on why slot machines, video games, apps, and devices create such an irresistible draw to continue pulling levers, leveling up, and refreshing screens. Social media, slot machines, and video games create "scarcity loops."

"Scarcity loops" - drive people or animals to repeat a harmful long-run behavior through short-term dopamine hits and satisfaction. A three-part process creates a scarcity loop:

1. **Opportunity**/potential to receive something of value
2. **Unpredictable rewards** -how much? When? An element of mystery!
3. **Quick Repeatability** -the more quickly you can repeat a behavior, *especially one with unpredictable rewards,* the brain becomes wired to repeat it again and again.

The consequences of scarcity loops go beyond habit. Cam Adair, a leading expert on video game addiction, paints a darker picture. Adair says, "Overexposure causes structural changes to your brain: numbed pleasure response, hyper-reactivity + willpower erosion." Adair adds, "Games are designed to keep you hooked using state-of-the-art behavioral psychology. They are fully immersive and provide dopamine overload."

Dopamine Hits

Dopamine isn't new to the modern world, and it's released in areas of our lives outside of device use. So why is it a big deal!? Why do people keep linking dopamine to negative things like the addictiveness of tech if we've all had dopamine hits before modern tech?

Our brains produce dopamine as a reward to motivate future behavior. We feel a dopamine rush after exercise, while eating delicious food, and during positive social interactions.

Dopamine evolved to encourage us to repeat behaviors that aid our own mental and physical well-being. Dopamine has evolved as a survival adaptation, motivating behaviors that are essential for personal survival and survival of the human species.

Addiction expert Nicholas Kardaras explains that natural dopaminergic activities "usually come only after effort and delay...but addictive drugs and addictive behaviors like gambling and (intentionally addictive apps and devices) provide a shortcut to this reward process which floods the nucleus accumbens with dopamine without serving a biological function."

Dopamine is a 'feel-good' chemical messenger (neurotransmitter) in the body. Our body likes how dopamine makes us feel, so it asks us to do more of what produces it. Without a biological benefit like nutrition, healthy exercise, or in-person social connection, we need to be wary of dopamine shortcuts.

Likes, comments, rewards, goal-meeting, and more flood our brains with dopamine. If we receive too much dopamine, we slowly devolve into a numbed pleasure response -the dopamine begins desensitizing our brains.

Dopamine overload:

- Challenges impulse control
- Hinders pro-social traits like empathy and compassion
- Fosters executive function decline (planning, organizing, prioritizing, etc)
- Creates an inability to focus and concentrate

Kardaras adds, "The more I stimulate (a person), the more I need to keep stimulating (them) in order to hold his or her attention. As with a drug addict, tolerance and desensitization develop, and the hyperstimulated (individual) needs ever-increasing levels of visual stimulation to stay engaged."

Do kids sometimes seem like zombies after screen time? Unable to focus on anything off-screen?

Embrace Boredom

Boredom sparks creative thinking and problem-solving. It flushes the mind and is part of the recipe for mental well-being. It's important to give our minds the space they need rather than filling the quiet with noise. Esteemed technology expert Cal Newport advises: "Embrace boredom. The broader point here is that the ability to concentrate is a skill that you have to train if you expect to do it well."

"A simple way to get started training this ability is to frequently expose yourself to boredom. If you instead always whip out your phone and bathe yourself in novel stimuli at the slightest hint of boredom, your brain will build a Pavlovian connection between boredom and stimuli, which means that when it comes time to think deeply about something (a boring task, at least in the sense that it lacks moment-to-moment novelty), your brain won't tolerate it," writes Newport.

Kids' Self-Report on Device Cravings

Tweens and teens have a love-hate relationship with their devices and apps. Several teens I know continue to feel drawn to apps like Instagram and TikTok weeks after removing them from their phones. Generational Researcher Jean Twenge asked her undergraduate students why they slept with their phones within reach. The interviews revealed, "Their answers were a profile in obsession." Students "needed" them for social media and videos before sleep and upon waking. They "needed" them in case they woke up in the middle of the night, like a modern-day teddy bear or special blanket. Twenge explains, "They talked about their phones the way an addict would talk about crack:

"I know I shouldn't, but I just can't help it."

"Having my phone closer to me while I'm sleeping is a comfort."

Journalist Nancy Jo Sales reports that the girls she interviewed simultaneously love and hate their devices and social media. Sales' thirteen-year-old interviewees say, "If I go on my phone to look at Snapchat, I go on it for like an hour, I lose track." Or, "I spend so much time on Instagram looking at other people's pictures, and sometimes I'll be like, 'Why am I spending time on this? And yet I keep doing it." These apps are so addictive that our kids cannot control themselves and often feel anxious about their inability to put them down.

Societal Tides

It's difficult for parents to stand their ground on screen time limits when societal tides flow in the other direction.

Children's Hospital Orange County illuminates the staggering number of hours kids spend on screens for entertainment, *unrelated to education.* As of 2024, teens spend over eight hours a day on devices outside of education, with tweens clocking 5.5 hours a day on average. The rise in screen usage directly correlates with the rise in adolescent anxiety, stress, and depression.

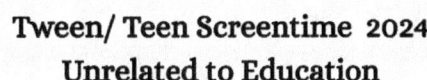

Tween/ Teen Screentime 2024
Unrelated to Education

Kids 8-12	5.5 Hours/ Day
Teens	8 Hours/Day

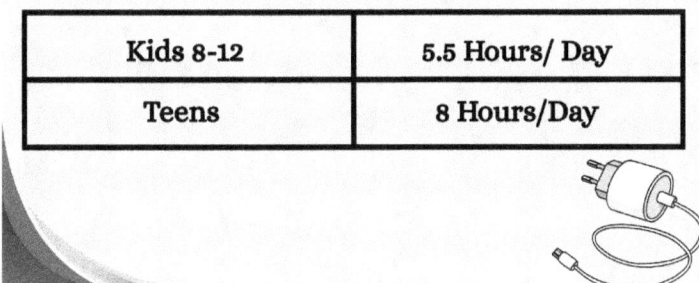

2024 CHOC DATA

It's deeply challenging for kids and parents when societal norms differ drastically from expert recommendations. Expert Randy Kuhlman writes, "Research found that kids who spend more than three hours a day playing video games reported less satisfaction, fewer social interactions, and more emotional problems than peers who played an hour per day."

It's hard to shift the parenting perspective from frustration to understanding. Device overuse drives me bananas, and I know I'm not alone. I'm trying to increase patience with kids and shift my frustration to tech companies whose profit margins are prioritized at the expense of collective consciousness and human well-being.

Tech companies built products optimized for engagement.

The intent of social media has shifted from increased belonging to a convincing enough imitation of it; pleasing shareholders replaced the original mission. It's never too late to try to pull kids further away from the jaws of tech -explaining the why behind the rules can help- and this chapter just gave you plenty of ammunition.

3

Chapter 3: Ideas for Addictive Screens

"It is okay to own a technology, what is not okay is to be owned by technology."

— Abhijit Naskar, Mucize Insan: When The World is Family

Kids 0-8

Families can create plans for the technology use they wish to model. Adults are drawn to devices -it's not their fault -they're designed to keep our attention. Children's brains are powerful sponges absorbing parents' habits through mirror neurons.

Cal Newport asked volunteers to do a 'digital declutter.' One parent shared his experience: "I started realizing how many (of his sons') small victories I miss out on because I feel this ridiculous need to check the news for the umpteenth time." The dad noted, "How surreal it can feel to be the only parent at the playground who is not looking down."

Avoid screens as pacifiers or babysitters. Babies and kids learn about the world on parents' errands, being at restaurants, and more. Exposure to different settings and situations helps children grow into the practice of entertaining themselves and processing new information. When screens keep young kids quiet on these occasions, it's much harder for them to engage socially as they get older. If toddlers are active or too noisy, walks can help!

Screens can easily become an irresistible crutch for care-givers -they're always on us, and they work. It's likely worth it to plan for the challenges that come with exposing kids to life outside the house. I oddly miss the days of endless Matchbox cars weighing down my purse. Wikkistix, coloring, and mini Playdough are good options.

Tweens, Teens, and Adults

Limit Kids' Screen Time

If families are confused, it is understandable! Experts caution against helicoptering and ask parents to give kids space to fail, learn, and grow. If parents limit screen time, are they interfering with kids' autonomy? Why are screens different? Like cigarette companies before them, tech companies use science to make apps, devices, games, and content consumption addictive.

Doctors do not recommend giving kids unlimited sugary and processed food and drink. Scientists do not recommend handing kids vast quantities of addictive drugs, alcohol, vapes, and cigarettes to let kids figure out limits and moderation. An endless supply of ingestible foods, drinks, and drugs could physically alter kids' brains and bodies. *Unlimited technology use can physically alter kids' brains and bodies.*

As noted in Chapter One, continual conversations help gain kids' buy-in to limits on screen time.

As kids enter their later teenage years, they will have to begin figuring out how to limit themselves. Ask them to take note of the time spent on their favorite apps every week and see if they are happy with the number. If not, what can they do about it?

Utilize Tools Within Apps and Devices

Apple and Android provide access to Google's tool, "Family Link." There are many other services out there; Family Link is the one we've used, and it's worked well. Device tools offer parents the ability to limit usage on particular apps as well as daily limits. Utilizing the tools within the devices helps to tone down the drama and battles between parents and kids. The device time limit ended the screen time, not the parent in the moment. This may minimize power struggles. Utilizing limits within the apps gives kids some of the autonomy they crave -they know how much time there is and they decide how and when they want to use it.

Device Rabbit Holes

Empower Kids to Utilize Modern Tools

Older teens and adults can set their own limits on app usage.

Ask kids to try an app like Rescue Time to track their time spent on technology. The results will likely be surprising. Forest is another fun choice for a rewards-based system for technology self-control. Several good options exist -the market has realized the need! Other choices include Block Site, and Opal.

Brick's popularity soared in 2025 and 2026 as a Millennial and Gen Z favorite for taking control of their tech lives. Brick is a "A physical device that temporarily removes distracting apps & their notifications from your phone. Designed for simplicity. No subscriptions. No complex setups. Just more time for what matters."

34

Continue Discussions with Young Adults (18-26)

Many high schools took a big leap in 2024 and 2025, banning classroom device use or full-day device use. Prior to 2024, many schools were without tech-use direction, and phones in kids' pockets exacerbated the magnetism between their minds and the devices. Young adult brains were at a vulnerable stage of development during the pandemic when most human interactions happened through technology.

Mel Robbins encourages adults to empower fellow adults who are struggling to find their way, to support and step back. Robbins refers to this as the "Let Them" theory. She does not suggest a full "Let Them" for young adults. "According to the experts, a human brain from a developmental perspective doesn't fully mature until age 25. Legally, you are an adult at 18, but from a neuroscience perspective, someone between the ages of 18 and 25 still needs a lot of guidance. So you have to be the adult in the room, getting professional help and steering what's happening."

Keep Technology Out of the Bedroom

When I read an expert recommend a **total void** of technology in kids' bedrooms, I thought it sounded like overkill. Then I read it from another expert... and another. The more I read, the more I realized that this is the advice of parenting experts and Silicon Valley technology executives alike. Even after establishing this guideline, I found that an old device was brought into the bedroom for an all-night Snapchat session. Continued conversations seemed to result in somewhat better choices.

Hopefully, enacting and discussing these measures with kids

under parents' roofs aids in creating good and lasting habits as kids get older. Physical alarm clock habits help to keep devices away from beds.

Indistractable

Nir Eyal's book, *Indistractable: How to Control Your Attention and Choose Your Life,* offers seemingly endless hacks to take control of our digital lives. It's a worthwhile read for kids and adults. In the book, Eyal quotes Richard Ryan (co-founder of the Self-Determination Theory) on the Needs Displacement Hypothesis. "The more you're not getting needs (autonomy, competence, and relatedness) satisfied in real life, reciprocally, the more you're going to get them satisfied in virtual realities." This can turn into a vicious cycle:

→Turn to screen time to satisfy needs

→ Real-life needs are not satisfied (time displacement)

→Turn back to screen use

Eyal's work reinforces a concept that runs through this entire book: intention and technology are like wind and a sail. Without both, you drift, directionless. Technology is powerful—but without intention guiding it, it won't move you or your family forward.

Eyal and Peter Gray are two of many who suggest that kids' lives lack free play, in-person connection time, screen-free time blocks, and downtime without screens. Ironically, we may need to schedule these things into kids' and adults' lives in this fast-paced world.

Watches and physical alarm clocks help us avoid constantly

checking devices for the time and becoming distracted. Scheduling technology use blocks help to keep tech use under control. "Without a clear plan, many kids are left to make impulsive decisions that often involve digital distraction," notes Eyal.

Family Phone Use Reflection

Kids aren't the only ones lured into the (intentionally designed) magnetic pull of tech. Grown-ups often realize a need to step back and gain awareness of when the phone creeps out at family dinners, watching shows together, and other times it disrupts connection. Kids absorb the habits that surround them, for better or worse. Family habits around device use may impact kids' future ability to regulate their own habits.

Building Barriers: Physical, Digital, and Creative

Physical and time barriers from tech help with the intentionality of use and can save us from falling down device rabbit holes. Kids will do better with these boundaries if we get their buy-in and their help in setting them.

Placing phones in another room -or even across the room- helps limit the constant desire to check. Timers keep us on track. A habit of airplane mode or silenced notifications outside of planned device time can give people hours of their lives back. Every time we get disrupted, it takes time to get back on track; studies suggest people are twenty-three minutes off course for every interruption to deep work. Notification muting has a profound impact on the magnetic pull of devices. Help kids disable autoplay and other features that don't have an end cue, if possible.

Under settings → display → color filters, phones can be turned to **grayscale**. This makes the visual of the phone less appealing and may act as a deterrent. If you make a chocolate cake and it tastes like Brussels sprouts, you're probably not going to devour it. Grayscale "diets" may work well, a January or Lenten challenge, a week here or there, or longer stretches.

In early 2026, some Gen Z folks took to making an "analog bag" to help keep themselves off their phones. The trend amplified, ironically, on TikTok. Akin to the purse full of Matchbox cars, toddlers, tweens, teens, and emerging adults may appreciate analog bags for device-free time. Ideas include card decks, hacky sacks, physical crossword puzzles, notebooks, mini football, knitting, Mad Libs, Mini Rubik's Cube, and a Magic 8-Ball. Exploding Kittens, Bananagrams, Mini Chess/Checkers, or Cards Against Humanity, depending on ages.

Nature Antidotes

Exercise used to be a natural part of life—hunting, gathering, farming, and other physically demanding activities were nec-essary for survival. We now add gym time and planned exercise into sedentary lives.

In decades past, our minds were less clogged with digital content. Today, planned extended nature time acts like an antidote to unclog and de-stress the mind. Many communities now offer nature play spaces that boost kids' creativity and problem-solving capabilities while fostering curiosity and a love of learning.

Train the Brain!

Our brains aren't muscles, but they respond to training in much the same way. If we train the brain to regularly cave to its urges to get on devices—out of habit, escapism, or other reasons—its ability to resist those impulses weakens. When we continually give in to the desire to be a couch potato, our muscles weaken. If we intentionally strengthen willpower by resisting the magnetic pull of devices and other temptations, we strengthen our mind's ability to resist.

Neuroscientist Andrew Huberman aims to resist impulses 20-30 times per day to strengthen his mind's willpower.

Why Am I on the Device?

If families are more intentional about device use, they gain more control over its addictive pull. Why am I scrolling? Am I reaching for my device/ switching tabs out of habit? To escape reality? To distract myself from stress? To procrastinate? Am I reaching for it robotically or intentionally? Is it controlling me or am I controlling it?

Outside Sources

Utilize outside sources for third-party validation of modern tech's addictiveness. Share the short 60 Minutes video, Brain Hacking, with kids. It offers an inside look at how tech companies manipulate people. Other options include The Social Dilemma, Childhood 2.0, and Social Animals.

Honest Conversations at Home

Admit to kids that technology use is tough for adults too - it is a challenge everyone faces. Have you ever felt sucked into a screen? Can you relate that it was hard to pull yourself away from something designed to pull you in? Maybe those mesmerizing typing dots when someone is responding to a message? Are you glued to emails, Instagram, or the news?

When kids see you on a screen, they don't always know why you're on it. I once took my kids on a work trip over Spring Break. After driving for seven hours, we grabbed a bite in the hotel restaurant at 10 pm. I pulled out my phone to check work emails and Slack messages that had accumulated while I was driving. My son mirrored my actions, pulled out his phone, and began to scroll. My first reaction was total annoyance, until I realized he didn't know I had dozens of work messages piling up and that I wasn't intentionally blowing off our "no screens at meals" rule. It would have been helpful if I had explained this.

Give kids positive encouragement and let them know that resisting the pull will be challenging, but they are capable. Try to resist blame. I used to play Candy Crush Saga and Candy Crush Soda Saga. I loved it -especially those bubbles in the Soda Saga! My son asked me why I don't play it anymore. It was hard to admit, but I felt like it started to control me instead of me controlling it. I didn't like that feeling. I felt this way after years of life experience. It's even tougher for kids with limited life experience to break away from screens that are intentionally addictive and all they have ever known.

When the Screen Goes Dark

Provide alternatives to screen time with mentally engaging activities: art, books, exercise, immersion in nature, writing — journaling and creative writing — building, learning a new language, cooking, social activities, creating a podcast, or designing a neighborhood scavenger hunt.

Discourage Multi-Tasking

Encourage kids and adults to be present when hanging out with others — watching shows, at meals, attending sporting events, and more. Being on devices during social situations exacerbates tech's addictive tendencies and diminishes connection.

Wait Until 8th

Brooke Shannon founded Wait Until 8th, a nonprofit that began as a grassroots organization in Austin, TX, in 2017. Over 130,000 families have signed a pledge vowing to wait until at least the end of eighth grade to give kids smartphones. The pledge remains anonymous until at least ten families from a school have signed it, then the organization connects the families and helps them spread the word. Wait Until 8th does not prohibit old-fashioned phones or smartwatches for calling and texting. By gaining traction within their community, families can fight against the societal tides.

Additional Resources for Technology Addiction

Experts disagree on the best approach — some favor a clean break, others a gradual reduction. Victoria Dunckley advises full abstinence from devices for at least four weeks to reset the nervous system. Nicholas Kardaras advises that weaning down works better than quitting cold turkey. You know your child best.

- Digital Detox, the Two-Week Tech Reset for Kids :Molly DeFrank
- Reset Your Child's Brain :Victoria Dunckley
- Glow Kids: How Screen Addiction Is Hijacking Our Kids - and How to Break the Trance :Nicholas Kardaras
- Cam Adair's video Video Game Addiction 101 for Parents & Therapists
- Cam Adair's website Game Quitters

The strategies in this chapter are not about perfection; they're about intention. Every small boundary, every honest conversation, every moment you choose connection over convenience is a step in the right direction. The sail is up. Harness the wind.

4

Chapter 4: Why Does Sleep Become a Parenting Struggle Again?

"The best bridge between despair and hope is a good night's sleep." -E. Joseph Cossman

Parents of newborns may envision a future in which sleep battles are a distant memory. Newborns often start with their days and nights mixed up. Once they have that sorted out, many babies morph into toddlers needing one more story, snack, glass of water, or a trip to the bathroom before bed. Toddlers often get up earlier than their parents would like.

Our son was *king* of the bedtime stalling tactics, which included "monster spray," and a slew of other requests. After a few years, these stages phase out, and parents think they are done with sleep battles. Until they hit the tween/teen years. If today's parents wished to stay out all night with friends (as a teen) they had to sneak out of the house. It was highly unlikely they would do this every night, and teens of decades past never had the temptation of being connected to friends 24 hours a day in their own beds. Being available to everyone all night is not the same as being close to them.

It is hard for parents to stand their ground while being accused of being out of touch, overly protective, and overly paranoid. Today's parents were all teenagers and many recall how out of touch their parents seemed. Believing parents are inept is often a genetic phase of development -families can't necessarily blame their kids for thinking this way. Today's kids are correct in that we are in uncharted territory. Kids are far more tech-savvy than their parents, who grew up with nothing remotely comparable.

Kids might be accurate when they tell parents that **ALL** their other friends are allowed to have screens in the bedroom. A National Institute of Health study found that 75% of American kids report keeping screens in the bedroom despite countless expert advice to the contrary. "The widespread use of portable electronic devices and the normalization of screen media devices in the bedroom is accompanied by a high prevalence of insufficient sleep, affecting a majority of adolescents (typically defined as ages 10-19), and 30% of toddlers, preschoolers, and school-age children."

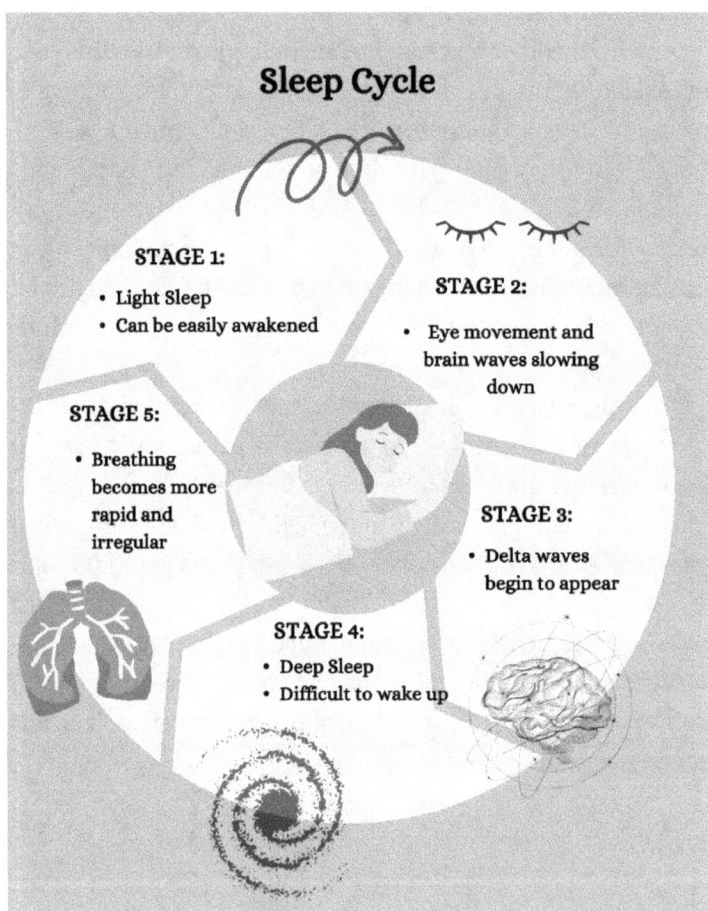

When early sleep stages are disrupted we lose much of the golden work sleep does for us in the later stages.

BBC News uncovered that more than half of kids report having a device within reach of their bed all night. Kids and parents are losing valuable sleep due to notifications. Even those with notifications *turned off* find it hard to pull away from the temptation of checking in on their device if they wake up. Surprisingly, even if people don't act upon looking at the device, the call of the temptation alone can interfere with sleep if technology is in the bedroom.

From the *Sleep Foundation*: "Repeated interruptions and awakenings can disrupt that process, causing far-reaching effects of disrupted sleep on brain function, physical health, and emotional well-being."

Sleep continuity throughout multiple stages and advancement into later sleep stages are crucial to the magic sleep works on our minds and bodies. **Technology in close proximity has a magnetic pull to disrupt sleep even when we're not actively using it.** *When sleep stages get interrupted the disruption pulls our bodies back to stage one.*

Impacts of Poor Sleep

1. Increased anxiety
2. Lower alertness
3. Decreased capacity to learn
4. Poorer memory formation and attention
5. Negative impact on emotional health and well-being
6. Impulse control issues, poor reaction times, and less focused concentration
7. Relationship stress, including greater conflicts
8. Lowered motivation to take care of oneself by eating right or exercising
9. Impaired immune system function

Matt Walker is a professor of neuroscience and psychology at the University of California Berkeley. In Walker's video, *Sleep Is Your Superpower*, he explains you need proper sleep to "Hit the save button on what you have learned. Without [proper] sleep your brain circuits become waterlogged and you can't absorb new memories." Walker mentions that when we lose one hour of sleep during daylight savings time, statistics show a 24% increase in heart attacks the following day, suggesting that even small amounts of sleep deprivation have significant effects on our cardiovascular system.

Screens are not only to blame when it comes to adolescents and inadequate sleep. Particularly, in middle and high school, kids' circadian rhythm shifts: they are physiologically wired to stay up later, but most schools create earlier start times for them. Kids are often overscheduled and carry a heavy academic burden.

Families can't control all the factors, but they can guide kids on bedroom technology use and bedtimes.

Kids often insist they need less sleep as they get older. Unfortunately, they're wrong. An NIH article explains, "Teens are in a time of very fast physical, intellectual, and emotional growth." Many need *more* sleep than they did at age ten and more than they will need as an adult. Michael Crocetti MD explains, "Teenagers are going through a second developmental stage of cognitive maturation," meaning they need more sleep. The World Health Organization and CDC recommend eight to ten hours. However, nearly 60% of middle schoolers and over 70% of high schoolers report getting less than the recommended amount of sleep on school nights.

The blue light in devices makes circadian rhythms even wackier. Chapter three mentions that experts recommend charging all devices in parents' bedrooms overnight to minimize temptation. Parents share that they have had to implement this rule as they watched the pull of the screen at night become irresistible. Kids are resourceful and often dig up old or spare devices to fill the void -keeping parents on their toes.

While tweens and teens are not likely to admit it, parental involvement can provide a welcome excuse not to partake in all-night Snap and Discord chats. They can blame their lame parents, and not have the pressure of having to stop the endless chat string themselves. Stanford Medicine calls teen sleep deprivation an epidemic.

The National Institute of Health informs that proper sleep "helps restore the brain by flushing out toxins that build up during waking hours." Kids' brains and bodies undergo massive developmental changes in the adolescent years. The ever-important nocturnal human growth hormone, which facilitates growth and cell regeneration, is "largely sleep-dependent," and adversely affected by poor sleep.

Getting through the stages of sleep properly can strengthen kids' immune systems and improve moods, concentration, organizational skills, fine motor skills, and memory. Scientists behind *The Benefits of Slumber* tout sleep's benefits for the cardiovascular system, breathing, and blood pressure, along with its assistance in problem-solving and attention to detail.

Ideas to Promote Better Sleep

1. No screens in the bedroom.
2. Journal thoughts and worries, or write to-do lists. Thoughts on paper get them out of our heads. Our brains feel reassured that the thoughts will be on paper tomorrow and not forgotten. Kids can keep a notebook and pen next to the bed to jot things down.
3. Calming activities such as meditation, relaxation techniques, reading a book, massage, visualization, and deep breathing.
4. Wind down individual screen use an hour before bed.
5. Creation of a good sleep environment, ideally cool, dark, and quiet.
6. Thoughtful eating and drinking so no one is too full nor too hungry. Experts suggest a small snack and a glass of milk before bed.
7. Ensure the bed and pillow are comfortable, with blanket weight that suits their preference.

8. Limit daytime napping to 20 minutes and never too close to bedtime.
9. Setting a bedtime that aligns with when they must wake up, and potentially allow for some wind-down time in bed.
10. Daytime physical activity, though not too close to bed-time.

The second sleep challenge often comes as a surprise to parents. But the work involved in coaching kids through it pays off—in happier, healthier, less anxious kids who are better equipped to face their days.

5

Chapter 5: Who Is at the Other End of Technology?

"A scam is a scam, a fraud is a fraud. Different rules don't apply in the City than they do for you or me." -Emily Thornberry

The 1980s White Van

Imagine dropping kids off in front of the proverbial white van—the creepy guy behind the wheel, inviting kids in, offering candy. Or driving kids to the mall in an unknown town and encouraging them to participate in fun kids' activities connected to porn halls and XXX movies. Unfortunately, most despicable child predators and shady characters have moved off the street and into families' houses, relocating to the internet.

Expert District Attorney Antoinette T Bacon explains, "Children are spending more time online, for school, for clubs, and for playdates. Parents don't know all the apps or how to use them, but sexual predators do. They know where the kids are and how to reach them. Just as parents taught kids to be safe at home by locking the doors at night, parents must learn how to keep kids safe online."

"Catfishing," reeling someone in through a fake online identity, got its name from a 2010 documentary by Nev Schulman. Schulman recounted how he finally met the love of his life through social media. However, he quickly learned that "Megan" and her stories regarding her and her family were a richly fabricated cast of fictional characters.

When people saw the film Catfish, Schulman received countless requests from people in online relationships begging him to help determine if their significant other was legit. Now, kids have become big targets for catfishers.

We lose crucial, non-verbal communication over the internet. This lack of proximity is a predator's best friend, and they know exactly what to say and do to gain a child's trust and become their so-called best friend. Professional groomers ask the right questions, lend a sympathetic ear, and compliment them. They use phrases such as "You sound really cool!" Or "What's your favorite movie/football team, YouTuber, or video game?" "Wait -you love Taylor Swift? Mr. Beast? Fortnite? Outer Banks? The Green Bay Packers?" They respond to kids' answers by saying, "No way -me too!" This enables them to bond with kids and gain their trust. Predators figure out what "type" of person kids want as their trusted confidante -male or female, age, race, hobbies -and easily "become" that person, pictures and all.

My friend Mary is an attorney on a trafficking task force. I asked her to describe the primary demographic of trafficked kids in the US. Mary's response: *Every* type of person from *every* type of background are among the trafficked. I was dumbfounded. Parents can't keep kids in a protective bubble from the scary, dangerous world, though they can try to prepare them. Children whose parents have spoken to them about online predators are less vulnerable than those whose parents have not talked to them.

As Mary reported, anyone can fall victim to a false identity online. In 2013, Notre Dame football player Manti Te'o was shocked to learn that his girlfriend, whom he had spoken to on the phone several times, was a false identity.

Predator Tactics:

1. Friending the target's contacts to appear legitimate
2. Engaging in increasingly personal conversations to build trust
3. Requesting to move conversations to other platforms or requesting pictures
4. Requesting offline contact
5. Targeting kids who give off vibes of loneliness, low self-esteem, or lack of engaged adult presence (particularly male role models)
6. Targeting kids who share publicly that they're going through a hard time -parent divorce, breakups, crushes
7. Reaching kids through their favorite games and platforms such as Xbox Live, iPad, computer, or phone
8. Using flattery and dangling opportunities: modeling offers, collabs (gaming, music, or other), money-making opportunities
9. Offering cheat codes, coupons, and advice within games
10. Turning kids against loved ones through validation ("I'm sorry your parents don't get you. I get you.")

Predators target kids through a host of apps such as Snapchat, Bumble, Instagram, TikTok, Tinder, What's App, Meet Me, Whisper, Ask.fm, Houseparty, Omegle, Kik, Holla, and Calculator% (a calculator app that secretly hides photos and messages). Kids may think parents are overprotective, but experts suggest initiating repeated conversations to be wary, particularly of "friends of friends" they encounter on social media.

There are countless stories of kids being duped online, often to horrifying outcomes. This happens to kids from *all* backgrounds and *all* levels of parental supervision.

A South Bend, IN, fifteen-year-old felt stressed and wanted to hang out with a friend. She reached out to her Snapchat friend list and let a friend of a friend pick her up. Little did she know he was a human trafficker.

More than 100 Philadelphia area teens, both boys and girls, felt compelled to send nude pictures or videos of themselves to a boy online, thinking they were communicating with their "friend," Haley.

When online predators receive even a mildly embarrassing picture, they often use it to blackmail kids by threatening to send it to their teachers, parents, friends, and even grandparents. Predators use this trick to convince the kids to send more and more embarrassing photos until they are in so deep that they feel completely helpless. These photos often end up on porn sites.

WHAT WE THOUGHT RAISING ADOLESCENTS WOULD BE LIKE 20 YEARS AGO

stranger danger = white van

that kid down the street might be a bad influence

turn that blasted music down

think for yourself! don't use Claude's paper or chat about the answers during the test

I wish he would stay home once in awhile! always out.

WHAT IT IS LIKE TODAY

stranger danger = in your home

764 or the manosphere might be a bad influence

I hope her image isn't deep faked

He might gamble his savings away in one night on his phone

think for yourself! don't use Claude or Chat

I miss connecting with you- can you take your earbuds out?

that dumb thing she did as a teen is on the internet forever

I wish he would go out once in awhile. always in his room.

By FBI estimates, there are 500,000 predators targeting kids online every day, each with a wide array of fake online profiles to choose from. More than 50% of victims are 12–15 years old, 89% are contacted in chat rooms and instant messaging, and 46% of kids give away information about themself online.

Widespread Peer Influence

As they raised Gen X kids, my parents only had to consider the influence that others in our suburb had on my brother and me. The world is kids' oyster now, for better or for worse.

Researchers Nicholas Christakis and James Fowler found that our peers influence our behavior to an incredible degree. Peers extend to the vast corners of the internet. We can find 'our people' on the internet to encourage just about anything, healthy or unhealthy. If we search for a topic such as 'cutting' or other self-harm, the algorithm will send us more posts on that topic. We can find strangers to befriend who will support exploration and action.

Radicalization leaders are among the strangers kids may befriend on the internet. Learn more from the NBC article *Teen terrorism inspired by social media is on the rise. Here's what we need to do.* Tweens and teens are at an awkward developmental stage -shedding some of their childhood identity on the path to discovering how they'll feel aligned as an adult. Adolescence is often a lonely and stressful experience, ripe for falling into communities where kids are made to feel important, accepted, and intelligent.

Adolescents are biologically wired to seek belonging—the emotional pain of exclusion is more powerful at this age. Online groups that bond over rage, self-harm, or disordered eating offer a shortcut to inclusion for kids desperately searching for their adult identity, and they rarely have those kids' best interests at heart.

59

The Hill reports that *Most of Gen Z are using TikTok for health advice.* A study of 1,000 young adults, ages 18-27, found that 56% of respondents use TikTok regularly for health and wellness advice, and 34% of those who do will not double-check the advice.

TikTok "therapists" are quickly rising in popularity, and some offer questionable advice with far-reaching consequences. Unlike licensed therapists who build context over time and understand the nuances of a family's dynamics, these creators deliver sweeping guidance to millions of strangers—and kids take it personally. Parent forums are filled with stories of heartbroken parents who have been abandoned by their children after following this kind of advice. In one post, a parent shared that their child's online "therapist" advised that parents who use Life 360 with their high schoolers are being abusive—and that estrangement was justified. Kids are choosing to cut off all ties with their families based on the recommendations of people who have never met them, don't know their families, and bear no professional accountability for the fallout.

The Parasocial Relationship

As a kid, Molly Ringwald was a central figure in my media. From her roles in John Hughes movies to her time on *The Facts of Life,* I was absorbed in every role. In 2001, I had the joy of sitting front row as Ringwald performed the starring role in Broadway's *Cabaret.* I was stunned when my mind instinctively felt like she might recognize me. From when I watched her from my living room? Twenty years prior? I joked about it at the time -why would my mind be so irrational?

I get why our kids feel attached to people they 'know' online, like Mr. Beast, Taylor Swift, or Andrew Tate. It must be incredibly difficult for kids to understand that their internet idols do not know them and may not have their best interests at heart.

The term "parasocial relationship" existed during my Molly Ringwald close encounter, but it has taken on a far greater significance as humanity's relationship with technology has further intertwined. Parasocial relationships are typically one-sided emotional attachments to another person, most often developed with a celebrity of sorts: an Influencer, YouTuber, Actor, Athlete, Singer, or more.

A friend of mine is gaining fame through activism and her burgeoning business. She was approached at a speaking event recently by someone who behaved as though they were close friends, though she had never met the woman. It was unsettling.

Parasocial relationships become even murkier because modern tech allows social media celebrities to direct message their followers automatically, fostering a seemingly closer connection. Many people, especially kids, could benefit from awareness of the parasocial relationship phenomenon.

The internet is a giant part of our lives. It can be hard to remember that people we 'know' through technology:

1. May not be who they say they are
2. May not be credible sources
3. Don't actually know us, or us them
4. Often do not have our best interests at heart

61

Kids are blissfully unaware of potential malice on the other end of their technology. The 80s Stranger Danger family conversations have moved from the White Van onto the internet.

6

Chapter 6: Conversation about Virtual Connections

"It is more important that innocence be protected than it is that guilt be punished, for guilt and crimes are so frequent in this world that they cannot all be punished." -John Adams

The central theme of Jonathan Haidt's wildly popular book, The Anxious Generation, is that kids are being overprotected and bubble-wrapped from real-world experiences while being under-protected from dangers and mental well-being threats in their virtual worlds. Communication and understanding of these dangers and threats could help kids fathom threats and parental concerns in the modern world.

Online Examples

A YouTube celebrity, Coby Persin, expressed concern about the prevalence of kids falling prey to online predators. Persin spoke with parents who were confident their child could not be duped and created a video to show families how easy it would be to lure a kid into becoming his friend on the internet. After this, he tested how quickly he could get them to meet him in person. The video is short and the results speak volumes: Coby Persin's The Dangers of Social Media (Child Predator Experiment).

Personal Examples

Do you know anyone who has been duped or nearly duped by people on the other end of the internet? I know a kind woman in her 70s who was romanced for 18 months by a 'London Financier' to the tune of most of her retirement savings.

My childhood next-door neighbor—let's call her Amy—is a college professor and published author. Not exactly someone you'd expect to fall for a scam.

Amy called up Apple Support after Googling the number. The woman on the other end of the call ('Barbara') was immensely helpful, spending 45 minutes helping Amy fix her malfunctioning AirPods. As they wrapped up, Barbara asked Amy if she had used her Apple Account in New Mexico recently. Amy had never been to New Mexico. Barbara gently mentioned that her account had purchased child pornography in New Mexico, and she may want to secure her account. Would she like to be transferred to her bank's fraud team? Of course, Amy gasped. After Barbara transferred the call, Amy's husband overheard Amy start to give away personal information, and his suspicion was aroused. He nudged Amy to ask the agent a few questions, and they realized she was being duped. The fraudsters had gotten a phony number onto the Search Engine, and Amy wasn't wary because *she had called the number.*

Kids and adults do NOT want to believe they are capable of being manipulated. Scammers are amazing at their 'jobs.' Tweens and teens may welcome warnings more gracefully if they hear real stories of multi-generational internet scam victims. There are many to choose from.

Virtual Stranger Danger

Remind kids that anyone they meet online could be forty-year-old 'Randy' even though their entire online persona looks like thirteen-year-old 'Isabella' or 'Caleb'.

Share How Predators Operate

Let kids know that 'Randy' is a highly trained professional and knows exactly how to sound just like 'Isabella.' He speaks teen girl lingo and is immersed in teen girl culture. He loves Olivia Rodrigo, Chappell Roan, and the latest season of Outer Banks. He uses the skull emoji 💀, meaning "I think it's funny even though I shouldn't," face with the hearts for eyes emoji , and words like "unc," "cringe," and "rizz," and speaks in ALL CAPS regularly.

Share that predators look for signs of loneliness, sadness, and vulnerability.

Chat about the way predators offer flattery, collaboration, or opportunity.

Confide that a predator will sometimes try to turn them against their family, or will offer validation - really 'getting' kids the way no one else does.

Friends of Friends

If you allow friends of friends, remind kids regularly that 'Randy' knows that friends of friends feel safe with just one degree of separation, and he has worked his way into the group.

NO In-Person Meetings

Never go to meet 'Isabella' or 'Caleb' (or anyone else you met online) in person.

Check Devices

Check in with kids and their devices, games, and websites regularly.

Privacy Settings

Set privacy settings to the max, no chat rooms with people they don't know. Let kids know that if they're asked to move conversations to another platform, it is a red flag.

Permissions

Required permission before downloading apps and games allows families the opportunity to research the games and be aware of the privacy and chat features.

Two-Factor Authentication

Enable Two-factor authentication on **ALL** kids' social media, games, and apps. It will seem like a pain, but if their account is hacked, it will be exponentially worse.

No Personal Info

Never give personal info such as address, phone number, or school name.

Usernames and Passwords

Usernames and passwords should not have any personal information.

No Photos Grandma Would Disapprove

Ask kids to **NEVER** send a photo of themself to anyone they wouldn't want their grandmother to see (or their principal, favorite teacher, pastor, etc). Let boys know it is never appropriate to ask for a compromising photo.

Encourage Family Communication

If someone sends or asks for an inappropriate photo, or personal info, or suggests meeting IRL (In Real Life), ask them to come to you.

The Parasocial Convo

Awareness of parasocial relationships helps our minds gain clarity on a confusing phenomenon. This term describes the feeling of being friends with someone we've never met, or the experience when someone feels like they know us personally when they don't.

Begin with the positive! Parasocial relationships are normal in today's world -they are not inherently harmful. They can provide inspiration for learning and motivation for personal growth. Communities around celebrities can foster true connection with like-minded fans. Connection with your favorite characters or entertainers can provide comfort and support through down days and tough times.

I had the pleasure of meeting the lovely Peloton instructor, Mariana Fernández, who has 56K Instagram Followers. She described the parasocial bond with her followers:

"It's a really unique dynamic because we have either messaged, or they've participated in classes that they've taken in their living room, so in a way, we are a part of people's homes, and by being part of their training or part of their movement, I feel like a lot of the walls break down, which encourages this kind of simpatico, this relationship that happens and that builds even if we never have met in person. So it's a really beautiful and unique way to form a bond, even when it hasn't been in person or we haven't spoken directly. But there's a connection that's made."

Challenges with parasocial relationships include:

· The illusion of intimacy: we feel close to someone, and

they do not feel that way about us
- Emotional investment imbalance
- Time lost to faux relationships could impact real-world connections

Thirty years ago, kids' answer to "What do you want to be when you grow up?" was often firefighter, doctor, or teacher. When asked today, it's more common for kids to answer with a vehicle to fame, such as a pop star, athlete, actor, or influencer. Share with kids that we often only see the shiny, filtered, edited side of fame. It has pros and cons.

Ideas for conversation:

"Who do you follow online that you feel connected to? What do you like about them?"

"Have you ever felt disappointed when someone you follow didn't live up to your expectations?"

"How do you think influencers decide what to share and what to keep private?"

"What's the difference between enjoying someone's content and feeling like you actually know them?"

Enjoying someone's content is normal. Following creators who inspire you is motivating. Real relationships—where both people know and care about each other—are what truly sustain us. In a world full of Randys and Isabellas, helping kids tell the difference might be one of the most important conversations we have.

7

Chapter 7: Why Do Kids Struggle with the Permanence of the Internet?

"The Internet is the first thing that humanity has built that humanity doesn't understand, the largest experiment in anarchy that we have ever had."

-Eric Schmidt (Former Google CEO)

There are many benefits to being a kid with modern tech. If I had access to Khan Academy, Crash Course, Coursera, and a host of other educational tools, who knows where I would have landed! My resources outside of the classroom or library rested upon our family's World Book Encyclopedia -stuck in 1980- claiming Yugoslavia to be a country and Pluto a planet. Would I trade our kids' lives with today's infinite resources for my life frozen in time with my 1980 information? I'm not sure. In the modern world, kids are saddled with anxiety, loss of innocence, and several other challenges that tech brings with it. This includes the Permanence of the Internet.

Kids are just kids. They say and do stupid things, and their silly antics are age-appropriate. Their brains continue to learn, grow, and develop until their mid-late twenties. Kids mature and learn by making mistakes. Even kids who have blossomed into wonderful, confident, intelligent beings will *still* make mistakes.

When friends in previous generations noticed mistakes, they may have ostracized, shunned, or even punched peers. But time healed. They were forgiven. They learned and grew. People they met in the future would typically only learn of past embarrassments and mistakes if the mistake-maker chose to talk about them.

Today's kids spend a significant amount of time communicating through the internet. If they make a mistake by writing their own words, creating their own images, or re-sharing others' words or images online, posts can be screenshotted and attached to their name for life.

Mistakes are easily uploaded, shared, re-shared, and potentially found by future employers, friends, and love interests. It is critical to communicate to kids to slow down, stop, and think before they hit enter.

It is difficult for everyone to envision their future life accurately.

We often think we are who we are, and there won't be much change. In fact, kids' opinions, interests, views, and even personalities will evolve and grow with experiences, friendships, and age. Their 25, 35, and 45-year-old selves will likely not appreciate certain TikTok challenges, Snapchat or Instagram posts, tweets, and screenshotted messages from friends and acquaintances of their teenage selves. This is especially true of personal information and compromising pictures.

Just as we were kids who did stupid things—because the emotional part of our brains was running the show long before the rational part caught up—today's kids are likely to do stupid things too." Unfortunately, their mistakes could be on the internet for all those in the future to see. There are no takebacks, do-overs, or delete buttons once a comment, picture, meme, joke, or article is posted. This *includes* comments within their friend chats. The internet is forever.

Ideas for the Permanence of the Internet

"If it doesn't matter, get rid of it. If you can't get rid of it, it matters." -Banksy

Open Dialogue

Calm reactions are key. A lack of anticipated blame or judgment over mistakes promotes honesty. Reassurance that parents will not judge, panic, or overreact lessens anxiety. Kids are relieved when they can lessen their burden of worry by coming to parents who don't amplify the drama. Approach the conversations with love. Let them know you are proud of them for coming to you!

Real-Life Examples

Watch episodes of *The Internet Ruined My Life* with kids

"This series tells stories of folks whose lives were ruined because they did something as simple as sending a tweet or posting a status update. Each episode features two people whose lives were turned upside down because of a few keystrokes."

Listen to the Hidden Brain Podcast, *You Can't Hit Unsend*

"Social media sites offer quick and easy ways to share ideas, crack jokes, find old friends. They can make us feel part of something big and wonderful and fast-moving. But the things we post don't go away. And they can come back to haunt us."

Delete the Post and Account ASAP

If kids make a mistake, delete the post right away and ideally delete the account. There is no guarantee, but experts highly recommend this course of action.

Think of it like SADD (Students Against Destructive Decisions), where kids can call parents to pick them up from a party with no shame or blame. The same principle applies here—if a child comes to you about a regrettable post, the priority is fixing it together, not assigning blame. Kids will make mistakes, learn, and grow. The goal is to make sure they feel safe enough to come to you before a bad situation gets worse.

Parents can sit down with kids and go through their messages and posts together. If parents find content kids should not have written or shared, kindly explain why it needs to come down—holding back the judgment they likely feel. This is much easier said than done. But communication lines remain far more open when parents manage to keep their agape reactions internal.

"Digital Citizen's Academy" advises: "Deleting the post is the better option, but deleting the account is the best option. The only way to truly remove an inappropriate post from social media is to delete the entire account. Remember, even if you delete a post from an account, the platform still holds all the account data on its servers, so as long as the person has an account, it can resurface. That information you thought was deleted can be flagged as relevant and anyone buying data from these platforms can see what was posted."

That last point is worth repeating to kids: deleting a post does not mean it's gone. As long as the account exists, the platform owns the data. This reality alone can be a powerful motivator for thinking before hitting send.

Become friends with your kids on all of their social media.

Regular Checks

Become friends with your kids on all of their social media. Frame this as a household policy from the start rather than a reaction to something going wrong -it's much easier to establish early than to introduce after a problem arises.

Sit down with kids once every month or two and look through their messages and posts together. Ask them what message they're trying to send. Do they think it could be taken the wrong way? These check-ins don't have to feel like audits; approach them with curiosity, and they become opportunities for conversation rather than confrontation.

A good rule of thumb: don't post or message anything you wouldn't want your grandma or school principal to see. Take it a step further...don't post anything you wouldn't want a future enemy to have access to. Kids might not have friction with anyone right now, but friendships shift, alliances change, and one never knows what the future will bring.

An alternate version for kids who respond better to scale: don't post anything you wouldn't want broadcast in front of your entire school with your name attached to it. Because it can happen at any time.

What Is Your Family Privacy Preference?

Opinions vary on kid phone privacy. Many families enact the rule that parents can look at kids' phones at any time - passwords must be given to parents and cannot be changed. This may act as a deterrent from messaging or posting inappropriate content. There is no black-and-white answer to family privacy preferences.

Abraham Lincoln Letter

Ask kids to channel their inner Abraham Lincoln. Lincoln amassed a large collection of unsent angry letters! He would write them to get his frustrations out on paper and put them aside. The New York Times explains, "Its purpose is twofold. It serves as a type of emotional catharsis, a way to let it all out without the repercussions of true engagement. And it acts as a strategic catharsis, an exercise in saying what you really think."

Writing helps us hone our focus to figure out what's really behind what we're feeling. Emotions can stir wild feelings and steer us away from the heart of the matter. This method helps kids practice slowing down to think before reacting.

Think Before You Hit Send

If Lincoln's approach sounds old-fashioned, the modern version is simple: type it out, then wait. Even ten minutes can change what you decide to send.

Consistent Vigilance

Kids' messages are only as private as the person they trust the least. Anything kids post can go public *at any time.* If someone wants to hurt them, they could do so within their online presence. Posts and messages can be copied and screenshotted, taken out of context, and then shared. How well do they know their friends and "friends of friends?"

Ongoing Conversations Remind Kids:

1. Respect and empathy for every post.
2. The permanence of online identity, which can affect their future college and job prospects.
3. Every meme, comment, joke, and picture people post shapes the identity with which others will see and judge them, now and well into the future.

Internet posts may live forever, as does parent support to navigate today's world.

8

Chapter 8: The Porn Chats

"It is quite reasonable to subscribe both to the old saw that no good girl was ever ruined by a book and to the perception that it is not good for children to be constantly exposed to the sexual violence in our popular culture. Protecting children seems to me logically, legally, and rather easily differentiated from censorship." -Molly Ivins

Kids, on average, stumble across porn on the internet between the ages of 8 and 11. They don't have to seek it out, it is specifically engineered to find them.

Many of today's parents grew up around Playboy magazines, and plenty of people have seen erotica in some form. There is a difference between previous decades and the world our kids live in now. In the past, those were choices made, and they had an endpoint. The magazine ended, the video ended, and the peep show ended. Today many kids don't have a choice: porn is perversely engineered to *find* kids, and the internet knows no end.

Many have experienced or watched a movie scene in which an accidental click morphs into a "not safe for work" moment that is impossible to click out of. This happens to kids in their own houses every day.

Psychologist Lisa Damour is regularly invited to speak at schools. A pastor at a Catholic School inquired if she would include porn in her speech for their assembly. The pastor explained, "I take confession from a lot of high school boys, and you would not believe how many of them tell me they watch porn late into the night. They feel horrible about it but they cannot get themselves to stop."

Times are different. Today's porn finds children and inspires a curiosity that can lead them to explore further. As discussed in Chapter Two, the body likes how dopamine feels - this feeling encourages people to do more of what produces it.

Dopamine motivates people to do things, again and again, creating addictions. Porn also provides those dopamine hits, and when combined with an endless supply, it often leads to more frequent viewing. Studies show more frequent viewing is correlated with negative perceptions of body image in both males and females.

Porn offers a faux connection. It creates physical sensations in the body, which makes it hard to distinguish porn from real intimacy.

Repetition causes desensitization, which is linked to sexual dysfunction. A report *Is Internet Pornography Causing Sexual Dysfunction?* indicates a sharp increase in sexual dysfunction rates under age forty and states that "Evidence has mounted that Internet pornography may be a factor in the rapid surge in rates of sexual dysfunction." The study shows that the percentage of males under forty with sexual dysfunction has skyrocketed to 26%. Porn use is associated with "diminished libido or erectile function."

Time Magazine shed light on the subject in *Porn and the Threat to Virility*. Psychology professor Philip Zimbardo explains, "Porn embeds you in what I call present hedonistic time zone," he says. "You seek pleasure and novelty and live for the moment." While not chemically addictive, he says, porn has the same effect on behavior as a drug addiction does: some people stop doing much else in favor of pursuing it. "And then the problem is, as you do this more and more, the reward centers of your brain lose the capacity for arousal." Zimbardo says that young brains and bodies may be particularly vulnerable to sexual dysfunction.

Research from "The Conversation", *Watching Pornography Rewires the Brain to a More Juvenile State*, explains how porn viewing by youth is linked to developmental problems. It has causational ties to depression and anxiety, is tied to lower quality of life, and alters brain chemistry.

"The other compelling finding in this study is that compulsive porn consumers find themselves wanting and needing more porn, even though they don't necessarily like it. This disconnect between wanting and liking is a "hallmark feature of reward circuitry dysregulation." When that happens, viewers often look for darker material, showcasing extreme violence against women.

A 17-year-old boy from Texas admitted to Jean Twenge, "I have never been in a relationship in my 17 years on this earth, and the big reason is porn and my association with it. At this point that makes me sad...Pornography, especially on the internet, has desensitized teens into not enjoying or wanting sex and intimacy."

79% of pornography viewing is done at home, says nonprofit Enough is Enough. Expert Donna Rice Hughes says, "Because of its impact on brains and behavior, many experts are calling it the drug of the new Millennium."

Singer Billie Eilish bravely spoke out on the *Howard Stern Show* about her experience with porn which began at age 11. She said "I think it really destroyed my brain. I felt incredibly devastated that I was exposed to so much porn...I think I had sleep paralysis and night terrors/ nightmares because of it."

Eilish continues, "The first few times I had sex I was not saying no to things that were not good because I thought that's what I was supposed to be attracted to." We need to help our girls and our boys understand this is not normal.

How Do Younger Kids Initially Find Porn?

1. Accidentally typing a wrong word or phrase.
2. Through friends, siblings, or adults.
3. Pranks at school or with friends
4. Accidentally clicking on a funny link, pop-up ad, or spam email.
5. Free game websites for kids can be hijacked with inappropriate versions.
6. Popular gaming consoles.

Savvy Cyber Kids cautions: "The restrained centerfolds and naughty XXX videos of yesterday have been replaced by hard-core, explicit performance-sex graphic imagery.... studies show that the most avid viewers of pornography are 12–17-year-olds. It is literally their de facto sex education."

Here is that tricky situation again: we want to talk to kids about things before they are exposed, but *without* sacrificing their innocence. Unfortunately, the world has changed to a place where experts suggest having these conversations earlier than we would like to. As uncomfortable as it might be, kids are better off learning about sex and pornography at home rather than on the internet.

Ideas for Conversation

Kids Under Eight

1. Teach the correct words for body parts
2. Let kids know they might accidentally find items that feel inappropriate on the internet. Ask them to come to you if they do
3. Collaborate with kids - develop strategies for potential situations. If someone shows them something they are uncomfortable with, they can say "I'll be right back," and get in touch with you

Kids 8-11

1. Continue teaching kids the correct words for body parts. Parents can expand anatomy knowledge at their level of comfort
2. Let kids know that they can talk to you about anything and continue to reinforce how to handle a situation that makes them feel uncomfortable. It may help to have a plan in place if a friend or peer shows kids something they

are uncomfortable with. Kids can say they need to use the restroom and call their parents

3. Don't allow your child to use the phone, tablet, or computer in their room alone. Keep electronic usage in the home in public family areas
4. Turning Teen is an excellent resource as kids enter the tween era

Kids 11-14

1. Acknowledge this may be an uncomfortable subject. Beginning with humor might help, "I used to think the sleepless nights of newborn parent life were hard, those were the days!"
2. Have conversations about sex without shame or judgment
3. Let kids know it's okay and normal to be curious about sex as you grow up
4. Share with kids that they can come to parents about anything, convos will be a safe space and they will never be in trouble
5. Ask questions such as:
   -Have you heard anything about it?
   -What do friends say about it?
   -Have they seen it?
   -Keep an open dialogue
6. Approach the conversation with love and be proud of them if they come to you!
7. Keep convos breezy and short, and aim for non-eye contact moments like driving or walking
8. Plan for *recurring* short breezy convos
9. Talk about sending pictures and sexting. Share that it is

illegal in many places to ask for, possess, or send pictures of private parts under the age of 18

10. De-stigmatize the fact that kids might be curious about sex as they go through puberty. Ask them not to use the internet for this curiosity. Perhaps offer kids a magazine if you are comfortable with that

11. Turning Teen continues to be a valuable resource

Kids 14-18

1. Real-world relationships mean consent and real interest. Coach teens that porn does not model sex while awake, sober, and in a healthy mental state. There should not be exploitation, blackmail, or abuse

2. Sex and relationships should be related to respect and trust

3. **Porn.is.not.real and it does not reflect the real world**

4. Much porn portrays violent acts against women that are not okay

5. Not everyone watches porn, and no one should feel pressured to watch it

6. Read the article: *Start here: Evolution has not prepared your brain for today's porn* and if you are comfortable share it with kids

7. Share with kids one of the Billie Eilish articles/videos: Billie Eilish on How Watching Porn When She Was Eleven Destroyed Her Brain
Billie Eilish Calls Porn a Disgrace and Destroys It

It may be understandably difficult for parents to believe their kids will be exposed to porn. Lisa Damour writes that "93% of boys and 62% of girls are exposed to it by age seventeen." The fact that the average age of exposure is between eight and eleven can be even harder to swallow. The modern world creates situations many parents find hard to fathom. Experts recommend family discussions around porn.

This topic was not on my initial parenting radar. There is plenty of public commentary blaming Billie Eilish's parents. I don't know anything about them, but I did read that she was homeschooled, and I would guess they were not absentee parents if this was the case. Unfortunately, this can happen to any of our kids.

There are numerous comments on the Billie Eilish articles from teenage boys, expressing despair that porn is ruining their lives and that they wish they could stop. We can't change the world they were born into. If we have enough open, honest communications with kids we can prepare them to handle this modern world with grace.

The Manosphere

Pornography exists within a broader culture of misogyny that targets boys.

The 2025 Netflix series *Adolescence* illuminated the rise of the 'Manosphere' (online misogyny) and 'Incel' (Involuntarily Celibate) culture popularized by celebrities like Andrew Tate. It's worth a watch. Parents may want to watch first and decide if their teens are ready for it. It can spark conversations about the radicalization of boys (as young as 11), inciting violence and hatred toward girls and women. It may offer insights into cyberbullying and the internet's impact on polarization. The BBC reports there was a 37% increase in violent crimes committed against women and girls between 2018 and 2023, and that violence against women and girls has reached 'epidemic levels.'

As touched on in Chapter Five, when technology and other factors exacerbate tween, teen, and adult loneliness, it's easier for personalities and members to draw kids into extreme groups like the manosphere. Teens in particular long for a place to belong, and supremacist or radical groups are all too ready to make them feel like a part of the community.

Adolescence creator Jack Thorne believes, "They say it takes a village to [raise] a child. It takes a village to destroy a child. If we're going to change the culture, there needs to be multiple solutions," he says. "We need everyone to lean into this problem to save these kids, to stop boys harming girls. It takes us all to do something."

9

Chapter 9: The Rise of Online Gambling

"You've got to know when to hold 'em, know when to fold 'em, know when to walk away, know when to run" - Kenny Rogers

Like porn's sibling, online gambling has addictive, dopamine-stimulating properties, lacks cues to stop, and is on a significant rise for adolescents and adults, negatively impacting the financial health of many. Like much on the internet, the ease of online gambling removes the barriers that existed as a deterrent. We no longer have to take a plane, car, or bus to gamble; we can gamble whenever and wherever we wish, for however long, often without friends or strangers in our corner, gently suggesting that we quit.

As mentioned in chapter two, when online gambling is not an option, residents who live within 10 miles of a casino are 90% more likely to have a gambling problem. Placing the casino inside our constant companion, our smartphone, amplifies the opportunity to make poor choices.

Bankruptcies have skyrocketed 25-30% since online gambling was legalized in the US in 2018, resulting in 30,000 more bankruptcies per year. A 2024 UCLA report found, "The legalization of sports gambling decreased consumer financial health. These results seem to be particularly pronounced when states legalize online betting, suggesting that the ease of access to gambling increases the problems associated with it."

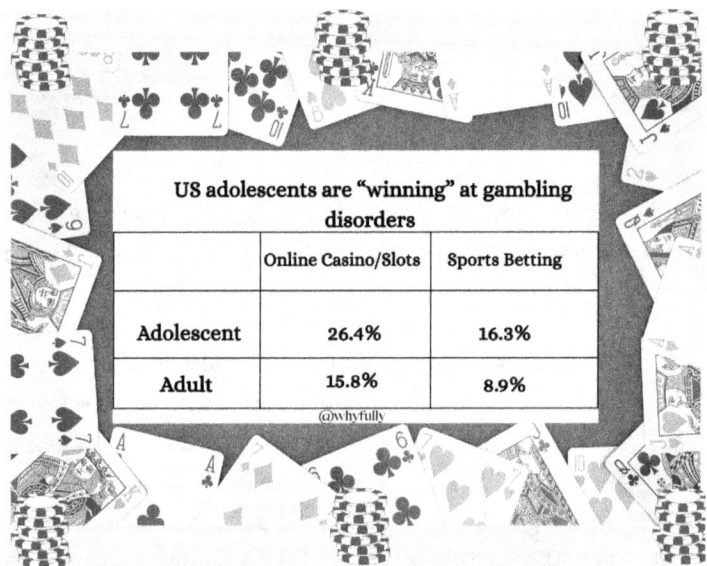

US adolescents are "winning" at gambling disorders		
	Online Casino/Slots	Sports Betting
Adolescent	26.4%	16.3%
Adult	15.8%	8.9%

@whyfully

Sources: CBS MORNING PLUS & THE LANCET (2024)

Expert Lia Nower cautions, "The groups most at risk are emerging adults and adolescents." Gambling firms get around age restrictions by framing bets as "sweepstakes." Adolescents have always been crafty at developing other workarounds for age limits!

60% of males on college campuses and many high school students partake in online gambling. Teens are far more likely to meet the definition of problem gambling than adults, and online gambling is pervasive in the adolescent world. If kids don't personally gamble online, they likely have friends who do. In a similar fashion to porn conversations, open chats work well:

1. What do you know about it?
2. Tell me about what is intriguing to you
3. How popular is online gambling at your school?
4. What ideas do you have about it?
5. Do you know how the odds are stacked? The house always wins—that's not a saying, it's a business model.
6. Real-life stories from gamblers whose lives have unraveled may be helpful

Acclaimed journalist Michael Lewis explored the rise of online gambling over a fascinating full season of his podcast (<u>Against the Rules: Fans)</u> in 2024-2025. The Supreme Court gave the green light to online gambling in the US in 2018. In 2017, when online gambling was illegal, legal bookies in the US cleared 300 million in revenue, and US consumers spent 5 billion on legal sports bets. In 2023, the industry scored an astounding 11 billion in profit, and consumers spent over 120 billion on betting, Lewis said.

The United States may be able to gaze into a crystal ball and get a glimpse of its future. Australia is 17 years ahead of the US in legalized online gambling. There are 26.6 million Australians. Australians lose 24 *billion* per year gambling. If the US follows suit, Americans will lose over $350 billion a year to the gambling industry.

Lewis digs into how predatory industries get away with exploitative practices, advertisements, misuse of behavioral psychology, manipulation of minors, and more. Historically, several industries have successfully exploited people for profit — tobacco, oil and gas, processed food, opioid/big pharma, lending, porn, gambling, and of course, modern tech. The playbook is remarkably consistent. How they do it:

1. **Exploit weak regulation.** Lewis says, "Strong market forces and weak regulations leave Americans exposed to predatory industries." A society that leans toward letting the market sort things out creates fertile ground for exploitation.

2. **Blame the victim.** Johnny has impulse control issues. Jack got hooked because he didn't know where else to turn. Olivia is lacking willpower. Emma is weak. Max doesn't know any better. Jenny was asking for trouble. Industries design their products to be addictive or harmful, then remain faceless by blaming the people who get addicted or affected, and offer "help" to pretend they care.

3. **Buy the science.** When people speak up about harmful practices, industries fund research to cast doubt on the evidence. The tobacco industry perfected this strategy for decades, and the same approach has been applied to climate change, processed food, and gambling.

4. **Buy the politicians.** Donations stall or subvert legislation through a practice known as regulatory capture -where the regulators end up serving the very industries they're

supposed to oversee.

5. **Buy the media.** Advertising dollars slow down the discussion of industry harms. Publications are less likely to run hard-hitting coverage of companies that fund their operations.

By the time society becomes concerned about the industry, we're in so deep it's difficult to find a way out. The gambling industry is betting—literally—that your family won't have these conversations. Proving them wrong starts with an open dialogue and a willingness to talk about money, odds, and impulse in ways most of us never had to growing up.

10

Chapter 10: Anxiety and Social Media

"To overcome the anxieties and depressions of contemporary life, individuals must become independent of the social environment to the degree that they no longer respond exclusively in terms of its rewards and punishments. To achieve such autonomy, a person has to learn to provide rewards to herself. She has to develop the ability to find enjoyment and purpose regardless of external circumstances."

— Mihaly Csikszentmihalyi, Flow: The Psychology of Optimal Experience

Why is anxiety, stress, and depression rising in kids today? How dramatic is the rise? Parents are not surprised that technology factors into the increase, but how specifically? Though families can't travel back to simpler times, they can arm themselves with information and uncover ideas to navigate these un-charted waters.

The pandemic of the early 2020s is often the scapegoat for kids' obsession with technology and tech's role in kids' anxiety. Before the global pandemic, psychologists were worried about year-over-year rising levels of anxiety and depression in kids.

A report from Georgetown University shares that depression and anxiety increased by 24% and 27%, respectively, from 2016 to 2019, *before* the added stress of the pandemic. There was a 21% increase in behavior problems in kids and a significant uptick in parent stress and anxiety during this period. Reports from Child Health Data indicate that one-third of all teens will have an anxiety disorder.

Atlantic editor Derek Thompson thinks American teen sad-ness is one of "the most interesting and disturbing mysteries in

America." Thompson says, "In the last decade-plus, the share of American High School students who say they feel persistent sadness has increased by 70% to the highest level ever recorded by the CDC. Today, more than four in 10 teenagers say they are consistently sad, hopeless, or anxious."

Tweens and teens' consistent device usage became the norm between 2010 and 2012. Usage grew continually year over year. The rise in usage directly correlates to the rise in adolescent anxiety.

Research, including <u>Jean Twenge's,</u> has established that the parallel between rising device use and rising anxiety is more than a coincidence. There is a causal link.

Chapters Ten through Fifteen illuminate the various ways modern tech contributes to anxiety and offer ideas to manage them.

Social Media and Traditional Media

Social Media's positives include connection with friends, inclusion in groups, and facilitation of communication. Though positives exist, social media leaves many people feeling lonelier than ever, particularly tween and teen girls.

Social media invites peer comparison, often resulting in kids' perceptions that they don't measure up. It spotlights negative news, sows division and polarization. It promotes unattainable standards of perfection through airbrushed filters and the common practice of posting only curated, ultra-happy moments. Social media brews loneliness that simmers or boils.

Jonathan Haidt writes that American teenagers, particularly girls, are reeling from the psychological effects of technology on a "massive scale" and that "the timing points to Social Media, particularly Instagram."

Why Young Brains Are Especially Vulnerable to Social Media sheds light on the strength of its impact on tweens and teens.

"Between the ages of 10 and 12, changes in the brain make social rewards—compliments on a new hairstyle, laughter from a classmate—start to feel a lot more satisfying..."

"Happy hormones' oxytocin and dopamine multiply in a part of the brain called the ventral striatum, making preteens extra sensitive to attention and admiration from others," writes Zara Abrams. Social media apps affect kids' brains in a different, and perhaps more intense, way than they affect adult brains.

Watching friends hang out without them induces tween and teen anxiety -perhaps friends post pics in real-time. Kids see their friends post about struggles and not talk about them. How do they know if there is a serious concern? What should they do? Kids feel anxious when they can see a message has been opened but not replied to. They might wonder if the person didn't like the message? Are they upset with them for some other reason? *How Tech and Social Media Are Making Us Feel Lonelier Than Ever* explains how being more connected through technology leads to greater isolation and loneliness.

Social media and traditional media are saturated with negative news. "If it bleeds, it leads." Cortisol levels (stress hormones) increase to anxious levels when we're regularly assaulted with negative news.

Combat Social Media Anxiety

Open dialogue strategies can work to lower social media anxiety. Conversations with kids about what they enjoy and don't care for on others' social media can spark thoughts into their own social media styles. Plan for chats outside of stressful moments.

Some social media posts show pride in accomplishments or are educational or informative. Others seem boastful and invite envy. Communication about how posts convey different messages helps kids navigate this terrain. I post about family trips primarily for the memories, and because it's an easy way to share experiences with friends I don't see often. I enjoy keeping up with others' adventures, and many enjoy mine.

As a culture and travel enthusiast, I find trip posts facilitate future conversations with friends in a way that describing events weeks or months after the fact could not do. The memories fade, and the descriptions become more vague. I make an effort to talk about appreciating others or offer insights in most of these posts.

Sit down with kids every few months.

What are they posting?

What message are they trying to send?

What do kids think of their friends' posts?

Are they trying to show off? Get attention? Share happy times?

How do they think others will feel or react when they see different posts?

If kids don't get flurries of likes or comments on their posts, will they be okay with that?

Parents can become friends with their kids on all their social media.

It's worth turning the lens on ourselves, too. Adults can take note of their own thoughts about social media posts and ask kids to do the same. A conversation about the analysis may spark deeper thought.

Do you feel happy for others when they post good things on social media?

Does it sometimes make you feel down?

Do you feel social comparison pressure?

Have you, as a parent, struggled with learning to manage your feelings about social media, or do you know anyone who has?

Do friends curate their posts to make it seem like they are happier than reality?

Share the article from apa.org with kids, *Why Young Brains Are Especially Vulnerable to Social Media.* It demystifies some developmental changes in their brain and explains how bigger emotions and stronger feelings are natural as they go through the tween/teen years.

Some parents opt to delay social media use as long as possible. Other families allow access solely from a home computer and do not give access to social media on portable devices. Tools exist to alert parents when apps are downloaded.

On the Plain English podcast, Haidt suggests that if people are on social media for more than half an hour a day, they try to cut usage in half or eliminate it entirely. Ask kids to try to limit checking of negative news.

There is bipartisan support in the U.S. to raise the age of social media to sixteen and to require age identification. Significant research suggests the medium is inappropriate and unhealthy for younger minds. In late 2025 and early 2026, Australia, Malaysia, and Portugal became the first countries to legislate social media age restrictions to sixteen and older. Much of Europe seems poised to follow suit.

11

Chapter 11: Information Overload • Attention Span

"The march of science and technology does not imply growing intellectual complexity in the lives of most people. It often means the opposite." -Thomas Sowell.

Information Overload

Friends and family comment that their memory seems to be faultier than it used to be. Our minds are too full!

Technology brings vast amounts of information, and our brains cannot process it properly, leaving us scattered and disrupting focus. The human brain evolved as a natural information filter, but when too much gunk is passed through a filter, it gets clogged.

*Daniel Levitin states that humans took in **five times** as much information in 2015 as they did in 1986!*

In a video, neuroscientist Daniel Levitin revealed that only 30 exabytes (a number followed by 18 zeros) of information existed in the world in 2012. By 2015, the world had amassed *300 exabytes of human-made information!* "We have created more information in the last few years than in all of human history before us, and we're assaulted by it every day."

In Derek Thompson's Plain English Podcast, *A Psychologist Explains Four Reasons the Internet Feels So Broken*, Thompson and Neuroscientist Jay Van Babel quantify the amount of news-feed the average person scrolls through on social media daily. Their analysis points to 300 feet -we take in approximately a 30-floor building, the Statue of Liberty, or a football field's worth of information from social media every day!

We have replaced conversation with content. One leaves you fuller, the other sends you back for more.

Frenetic Task Switching

Our brains aren't just overwhelmed by the volume of information — they're also exhausted by how quickly we're expected to bounce between it.

If asked to picture minds at their calmest, we may envision a relaxed state after a massage, chilling at the beach, or immersing ourselves in nature. Minds in a state of peace or serene focus.

Visions of an anxious mind likely include rapid thoughts and a frenetic state. The lyrics of U2's *"Until the End of the World"* might play, "In my dream, I was drowning my sorrows, but my sorrows, they learned to swim. Surrounding me, going down on me, spilling over the brim."

Kids' proclivities toward rapid task switching on devices are one of the culprits in the rise of their anxiety and stress. Research shows that the relationship between rapid task switching and anxiety is a two-way street. People task-switch when they're anxious and stressed, and task-switching creates stress. Digital media authority Gloria Mark notes that rapid task switching itself causes and exacerbates stress, resulting in physiological symptoms such as raised blood pressure.

Researchers studied how often students switched tabs on devices in 2014. "Results showed that switches occurred every *19 seconds.*" This was over a decade ago -the pace is only intensifying. Gloria Mark observed that task-switching frequency is on the rise as our world gets busier and busier. The frenetic activity in kids' brains is likely at a fever pitch.

Daniel Levitin explains that when multitasking, we "fractionate our attention into little bits and pieces, not really fully engaging in any one thing." This depletes the brain's energy and erodes the neurochemicals needed for focus. Levitin says, "Multitasking has been found to increase the production of the stress hormone cortisol, as well as the fight-or-flight hormone adrenaline, which can overstress your brain and cause mental fog or scrambled thinking."

"Multitasking creates a dopamine-addiction feedback loop, effectively rewarding the brain for losing focus and constantly searching for external stimulation..."

"It is the ultimate empty-calorie brain candy. Instead of reaping the big rewards which come from sustained, focused effort, we instead reap empty rewards from completing a thousand little sugarcoated tasks." He adds that constantly checking social media, email, the news, and Twitter can "constitute a neural addiction."

Attention Span

Attention Span on Devices

2003	150 Seconds
2012	75 Seconds
2023	47 Seconds

Research from Gloria Mark measured human attention span on devices over twenty years

Brain Rot was Oxford Dictionary's Choice for the 2024 word of the year. The phrase 'Brain Rot' is "Often invoked by young people on social media to describe the 'supposed deterioration of a person's mental or intellectual state,' particularly stemming from overconsumption of trivial online content," writes

Jennifer Schuessler.

Reading on technology trains the brain to "T" read, "F" read, or "Z" read. We read the top and then scan the content, losing the ability to deeply absorb the writing. Literacy researcher Maryanne Wolf voices concerns about the changes digital reading makes to our brains and how it diminishes creative thinking. When we skim, "We don't have time to grasp complexity, to understand another's feelings, to perceive beauty, and to create thoughts of the reader's own," writes Wolf. Studies show that reading physical books enhances memory and understanding of the subject matter.

Short-form creations from TikTok, Instagram, and YouTube are impacting attention spans. Computer scientist Nyel Ansari notes that we tend to have FOMO (Fear of Missing Out) on the latest information, which propels us to seek out and skim as much content as possible. Ansari says, "We prioritize quantity over quality, speed over depth...we're past the point of consuming information. We're at the point where it consumes us." Ayel found himself falling prey to short-form content over-consumption. He noticed that excessive short-form content destroyed his focus, presence, and ability to engage with the world around him. Ansari suggests adding friction to our app use by logging out every time we use apps that provide dopamine hits.

Overconsumption of short-form content can reshape neural pathways to crave more instant dopamine hits, weakening focus and strengthening instant gratification craving. We train our brains to crave shorter and shorter form content. When we become accustomed to reels, TikToks, and YouTube Shorts, longer, deeper content seems difficult or boring.

Amplify Focus

"Sometimes, success is less about making the good habits easy, and more about making the bad habits difficult." –James Clear

Timers work well for limiting passive scrolling and keeping device use intentional

- Envision Brain Rot when consuming online content
- Set timers or limits on digital consumption -most kids (and

adults) don't realize how much time they lose to passive consumption
- Apps like RescueTime can help with a more accurate picture
- Prioritize reading and aim for physical books over digital ones
- Ask kids to try to read for longer stretches, do puzzles, Legos, or other creative activities for extended periods to hone patience and attention spans
- Literacy expert Maryanne Wolf advises that children ages 0-10 only read in print to develop deeper reading skills
- After age ten, Wolf advises kids to keep print in the mix if they opt to read some digital content
- Actively limit reels, YouTube Shorts, TikToks, and other short-form content to enhance attention span and focus
- **Log out** of apps that tempt endless scrolling or short-form content - the hassle of logging back in may be a helpful deterrent
- 45-60 days of daily meditation (20 mins) can have significant impacts on memory and focus, and can lower anxiety

Our attention is not infinite -it's one of the most valuable resources we have, and right now it's being spent faster than we realize. Every scroll, every tab switch, every fifteen-second video trains our brains to expect more, faster, shorter. The good news is that the brain is remarkably adaptable. The same plasticity that allows it to be rewired by short-form content also allows it to be rewired back -through reading, focus, boredom, and presence. Modeling these habits at home might be one of the most powerful things a parent can do.

12

Chapter 12: The Time Displacement of Well-Being Activities

"A certain amount of reverie is good, like a narcotic in discreet doses. It soothes the fever, occasionally high, of the brain at work, and produces in the mind a soft, fresh vapor that corrects the all too angular contours of pure thought, fills up the gaps and intervals here and there, binds them together, and dulls the sharp corners of ideas. But too much reverie submerges and drowns. Thought is the labor of the intellect, reverie is its pleasure. To replace thought with reverie is to confound poison with nourishment." -Victor Hugo

Sleep

As noted in Chapter Four, technology is displacing kids' sleep and diminishing well-being. Stanford Medicine declares, *Among Teens, Sleep Deprivation Is an Epidemic.* Inadequate sleep impacts alertness, memory formation, attention, emotional health and well-being, impulse control, and immune systems, and increases anxiety. Netflix CEO Reed Hastings fully admits that sleep is their biggest competition and brags that his company is winning.

Peers pressure tweens and teens to be available to "chat" at all hours of the day and night. Peer expectations lead to insufficient sleep, throwing off the body's homeostasis (balance). Waiting for responses and expectations to respond causes anxiety -for example, waiting for mesmerizing typing dots....

The Happiness Activities Screens Displace

Technology displaces activities that spark joy and well-being, such as creative thinking, and good old-fashioned connection through hanging out with friends. Interpersonal connection and communication are vital to thriving and flourishing. With adults and kids turning to their phones to minimize potential awkwardness, we close off conversation opportunities and diminish human connection.

Kids and adults used to spend **considerably** more time reading, relaxing, exercising, creating, unwinding, and doing other activities that helped build health and well-being.

Technology overrides invaluable downtime that fosters creative thinking and problem-solving, allowing our brains to sort through thoughts and events of the day. Downtime without screens acts like a vinegar and baking soda remedy for our clogged mental filters. Awake time *without distraction* is a necessary component of mental well-being.

Jean Twenge cites data showing Gen Z adolescents and young adults spend less time on homework, paid work, volunteering, and extracurricular activities. Where does the extra time go? *To devices.* She references multiple studies that show direct correlations and causations between personal screen time and increased unhappiness. Twenge warns, "With teens spending less time on activities that assuage loneliness, and more time on those that don't, it is not surprising that loneliness has increased."

Through data analysis, Twenge found that college students in the 2010s spent *seven fewer hours per week* socializing with friends than college students of the late 1980s. The dramatic drop in in-person connection mirrors the rise in anxiety, stress, and depression. Twenge notes this is time lost on "Building social skills, negotiating relationships, and navigating emotions."

Gen Z and their younger counterparts are on track to spend 25 years of their lives on mobile devices.

An Alluring Virtual World

Tech's capability of amplifying the senses may dampen our enthrallment with the real world. The Northern Lights look more stunning on my phone than with the naked eye. I notice details in pictures I take that I missed when looking without my device. I am a huge fan of the fictional book *Ready Player One: A Novel,* a dystopian adventure in which the virtual world is a far more interesting place to be than the real world. Its eerie plausibility is part of the draw. Who wants to deal with real-world chores, homework, practice, bills, and other monotonous parts of life when there is a mesmerizing place to escape to in our pocket?

Gaming

Video games are an opportunity for kids to escape real-world situations and feelings. In short increments, they often serve as stress-relieving diversions or ways to connect with friends. In longer time spans video games may fuel anxiety and create and deepen habits of avoidance which hampers resilience and distress tolerance. The games may become addictive and contribute to or cause social anxiety, a condition in which people are more comfortable living virtually than in the real world. Overuse of video games may lead to decreased social skills, impulse control issues, decline in executive function, and poor concentration.

A state of 'flow,' coined by Mihaly Csikszentmihalyi, is correlated with greater happiness and life satisfaction. Flow involves pushing through challenges and obstacles on the path to achievement and accomplishment. When in a state of flow, people get lost inside their (real-world) activity and become intensely focused. Kids and adults gain strength and confidence when they master things and experience "flow." Csikszentmihalyi explains, "The best moments in our lives, are not the passive, receptive, relaxing times—although such experiences can also be enjoyable if we have worked hard to attain them. The best moments usually occur when a person's body or mind is stretched to its limits in a voluntary effort to accomplish something difficult and worthwhile."

Kids might experience flow when building Legos, playing sports, writing, drawing, creating art, getting lost in an intriguing book, cooking, and more. To get to the level of mastery or flow, kids need to fail repeatedly, feel down about it, keep trying, and eventually bask in the warmth of a successful flow state.

Video games give us a "faux flow." Bodies experience the physical rush of flow by pushing through real-life challenges and obstacles on a path to achievement and accomplishment –our minds really like this!

True flow comes from accomplishment in the real world, creating or accomplishing something tangible. When we finish our level, catch the Pokemon, win the game/race, slay the zombie, or steal the car, the flow we felt in the moment gets replaced by reality –our success was not real. This realization may encourage escapism from reality by turning back to the game world.

Nicholas Kardaras writes, "Kids have magical fantasy worlds to lose and reinvent themselves in; worlds where they can create strong and powerful majestic personas that get to shoot everyone into oblivion, all while pursuing some noble common goal."

Video games flood kids' brains with dopamine, which can lead to desensitization, particularly in brains that are in their development phase until age 26. Moderation isn't just preferable, it's essential.

Rekindle Sleep and Well-Being Activities

"The results could not be clearer...All screen activities are linked to less happiness. All non-screen activities are linked to more happiness. Screen time (particularly social media use) does indeed cause unhappiness." -Jean Twenge

Proper sleep is imperative to kids' physical growth and health. Sleep aids learning, memory formation, mental well-being, and more. Modern technology displaces many kids' proper sleep time.

Kids can balance the consuming nature of screen time with the brain-boosting act of creating. Reading is a fantastic activity; the reader creates visual scenes in their heads while reading the author's words. Creative writing, journaling, photography, drawing, painting or other art, cooking, and learning a new language are brain-boosting activities. These pursuits build confidence and self-worth and feed our souls.

Encourage soul-filling activities before screen time when kids' brains are not already sapped by the consumption and dopamine overload that screen time can bring.

Create space: treat yourself to solitude!

"Isolation is aloneness that feels forced upon you, like a punishment. Solitude is aloneness you choose and embrace. I think great things can come out of solitude, out of going to a place where all is quiet except the beating of your heart." – Jeanne Marie Laskas

Peace in the Present

When we work to stop filling in the cracks of our free time with device use, we strengthen impulse control, expand patience, and benefit from staving off urges for instant gratification

Actively taking in our surroundings -what we can see, smell, hear, and feel –is a therapeutic practice to promote calm and well-being. One simple practice psychologists recommend is grounding ourselves in the present through our senses:

Five things we can see

Four things we can touch

Three things we can hear

Two things we can smell

One thing we can taste

Before smartphones and tablets, we did this regularly –we had little else to occupy us. We can work to revive this practice by challenging ourselves (parents and kids) to be in public, car rides, bus rides, and more without filling the time with device use. This doesn't mean we can't look at devices at all; we can try looking for particular reasons and then putting modern tech away. It's harder than I would have thought, particularly when phones fill the hands of those in our vicinity. It may feel like wearing a giant magnet while standing near a metal object. We may feel the physical, powerful draw and strengthen our muscles to hold our ground.

Screens displace kids' boredom, which is healthy for development. Do you remember as a kid saying, "I'm bored!?" It turns out that boredom is good for our brains! Like its cousin, sleep, boredom helps us process, organize, and store information. Boredom boosts creative thinking and problem–solving. As Cal

Newport says, "Embrace boredom!"

Confidence Comes from Real-World Accomplishments

Setting goals and actualizing accomplishments aid in well-being. If kids spend the day glued to a Snapchat feed, refreshing for Insta likes, battling Fortnite enemies, scrolling TikTok, or glued to YouTube videos, it impedes their well-being from real-world accomplishments. Lack of real-world fulfillment erodes confidence and creates a sense of hollowness.

'Satisfaction' on screens removes some of the work involved in real-world relationships and accomplishments. Screens mimic flow and connection well enough that we stop reaching for the real things.

Level Up on Gaming Time

Limits on video games may enhance current and future impulse control. Kids and adults easily lose track of time when they are engrossed in a virtual world that is designed to be more exciting than their real one. Add non-virtual activities like reading or creative pursuits to kids' free time. Enable opportunities for in-person peer or friend time. Hikes or beach visits connect kids with nature and give bodies and minds a reset. Bring or send kids on errands like the grocery store or Target runs to foster human interaction.

Teach kids about flow. Help kids understand why the rush of the video game feels incredible in the moment but may leave them feeling empty later, like 'faux flow.' Kids might feel the need to fill the emptiness by returning to the rush they get from the game. Encourage kids to seek real-life flow feelings from creating, writing, athletics, in-person friend connections, or other achievements.

Encourage screen-free experiences with friends. I hear people comment that gaming together is merely the new way of playing together -there's communal activity -it's social -it shouldn't matter that kids are together on screens -they're still together!

This didn't feel right to me, but I couldn't think of a way to explain why it was different, other than that it often contributed to individuation by putting them on individual devices. Thankfully, Jonathan Haidt had an answer in *The Anxious Generation*. Haidt explained that when kids play together without devices, they experience personal growth from creating the parameters of their free play with each other. This strengthens skills of cooperation, conflict resolution, problem-solving, creativity, and metacognition in a way that playing a game within video games' pre-established (and often unchangeable) parameters cannot.

Every hour spent on a screen is an hour not spent on something else. That's not a guilt trip -it's just math. The goal isn't to eliminate screen time but to make sure it isn't subtly crowding out the sleep, play, boredom, and connection that kids need to thrive.

13

Chapter 13: Individuation and Loss of Autonomy

"Each of us is now electronically connected to the globe, and yet we feel utterly alone." – Robert Langdon, Angels and Demons by Dan Brown

A Rise in Individualization and Fragmenting Society

Growing up, I was certain that once everyone accepted one another's differences, stress and anxiety would decrease. It turns out that being a part of a group at least slightly blunts the impact of stressors done to the group. We feel camaraderie when we're with our people. When we're attacked as an individual, anxiety usually hits us harder. Accepting others' differences is profoundly valuable, and seeking commonalities while accepting differences becomes equally important to mental well-being.

Individuality creates a stronger attachment to personal feelings. When we identify primarily as individuals rather than as part of a group, every setback, criticism, or stressor lands squarely on us. There's no team to absorb the blow. When we belong to something bigger -a community, a team, a close-knit group of friends- stress gets distributed. A bad day feels more manageable when it's shared. Groups blunt the impact of stressful feelings in a way that scrolling alone on a device never can.

I remember the depths of isolation when I felt alone in a particular belief. I felt confident that there were no weapons of mass destruction in Iraq in 2002. Though a handful of people agreed, they were not in my orbit. Everyone, no matter their left, right, or center leanings, told me that the idea was bananas. When the truth came out (that there were no weapons of mass destruction), the news cycle had moved on, and it was brushed off. I was alone in my outrage, leading to a decision to pay far less attention to the news du jour.

Technology intensifies individuality and isolation. Today, algorithmic personalization creates unique information bubbles. Netflix recommendations, news feed, and search results are individually tailored, reinforcing existing preferences and gradually separating the communal media experience. Shared cultural moments become rarer.

Modern tech enables extreme customization of identity and lifestyle. One can curate their online presence, find niche communities for any interest, and construct a highly individualized worldview. We're increasingly living in personalized realities rather than shared ones, which sometimes feels liberating in the short term.

Fewer commonalities to celebrate, commiserate, or experience lead to greater isolation and loneliness. Friends used to watch the same shows at the same time, sparking animated discussions the next day at the water cooler, at recess, or in the student union. Today we have thousands of friends and feel less connection. The staggering number of shows, movies, podcasts, YouTube videos, and games available on demand has changed when, where, and what kids watch, limiting communal discussion opportunities. The world is fracturing further and further, and technology shoulders much of the blame.

Technology has sown more division than connection. Jonathan Haidt's Atlantic article, *Why the Past Ten Years of American Life Have Been Uniquely Stupid* became one of the publication's most-read pieces and is an enlightening deep dive into how technology has sown division. Political and social polarization further alienates people, begetting individuation and loneliness. Fake news, bots, and deep fakes sow division and leave people wondering where reality begins and where misinformation ends. This makes our minds uncomfortable.

Lack of Control and Autonomy

Technology sneakily snatches autonomy and a sense of control over our lives. Our brains like the accomplishment and autonomy that come with turning a page, checking things off a physical list, and knowing where we are and where we're going on a physical map. Studies show our brains will absorb, comprehend, and retain information better from physical books, notebooks, planners, and more.

Google Maps helps us avoid real-time traffic, but ceding mental knowledge and problem-solving to machines inadvertently diminishes well-being. When we outsource navigation, we lose the inner confidence that comes from knowing where we are in the world. That sense of agency extends far beyond driving directions. The future of AI will continue to throw our autonomy out of whack as we surrender control of memory, attention, focus, interpersonal connection opportunities, and navigational competence to devices.

Hopelessness/Overwhelm Over Its Addictive Properties

Addiction takes a toll on well-being when a substance -or behavior- controls us. Many of modern technology's apps, games, devices, and media outlets have been manipulated by behavioral psychologists to keep our attention. Technology's addictive properties are challenging for many adults and wreak havoc on interpersonal relationships. Kids who haven't experienced life before modern tech may not realize the power of its addictive properties.

Addiction, anxiety, and depression are inextricably linked. Addiction ramps up anxiety and depression, and anxiety and depression beget addiction. Technology addiction exacerbates patterns of choosing what feels good in the moment over better choices for the long term. Excessive dopamine from technology triggers cortisol production, releasing stress hormones. The dopamine overload kids get from modern technology promotes and heightens impulse control challenges.

Build Community

The American ideal of individuality is one of the reasons we feel more isolated and alone than previous generations. The other main culprit is technology, which provides exponentially growing information and choice, which amplifies individualistic feelings.

With these advancements come stress and anxiety -we feel less and less connected than previous generations did. When the modern world diminishes feelings of connection, families may want to be more intentional about building connection and community back into kids' lives.

Connection Is the Antidote: Four Ways to Build It

Free Play

Unstructured outdoor play builds creativity, negotiation, and independence. Mixed-age neighborhood play is especially powerful -and increasingly rare.

Ensemble & Arts

Theater, choir, and band build shared purpose and belonging while fostering empathy and perspective-taking.

Team Sports

Shared wins and losses create belonging. Being part of a team distributes stress in ways scrolling alone never can.

Community Projects

Working toward something bigger than yourself builds empathy, purpose, and connection to others.

whyfully

Honing skills of empathy and pro-social behavior builds connection. Acts of gratitude and kindness help kids feel more connected to others and the world. Immersion in nature and visits to large bodies of water foster a connection to the world.

Aiming for tech use on shared screens versus individual

devices mitigates some of the individuation. Watching shows together can spark conversation and togetherness.

Intentionality with free play within local communities could spur mixed-age play. Mixed-age play has proven benefits and was historically the norm. I have fond memories of hours spent with neighborhood boys and girls whose ages spanned across seven years on Wilson Drive. Younger kids gain information and knowledge in a different way than when playing in mixed-aged groups. Older kids receive benefits of mentorship, leadership, and different skills of cooperation and problem-solving. Mixed-age play enhances traits such as empathy and perspective-taking.

My son is great about voicing his requests to schedule time with lifelong friends or family groups -and with hectic lives, I'm thankful for the reminder! I'm going to try to set bi-annual reminders to go through our lists of these groups and try to get something on the books.

RECLAIM YOUR MIND

PHYSICAL BOOKS

Our minds absorb, retain, and comprehend information better from physical pages than screens.

PAPER PLANNING

Physical planners give minds the satisfaction of control and accomplishment that digital tools subtly take away.

PRINTED MAPS

Knowing where you are in the world builds inner confidence and a sense of agency that GPS erodes.

HANDWRITTEN NOTES

Handwriting engages the brain differently than typing -improving memory, comprehension, and creative thinking.

CASH & LEDGERS

 whyfully

Physically handling money builds a stronger grasp of financial reality than tapping a screen ever can.

JOURNALS

Writing by hand slows the mind down in a healthy way - building self-awareness, reflection, and emotional clarity.

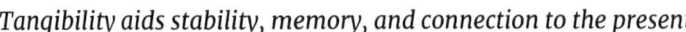

Tangibility aids stability, memory, and connection to the present

Give Minds Control

Tweens and teens have far less autonomy than their parents did, and this is considered to be the other major contributing factor in the rise of adolescent anxiety. Technology is one of many forces eroding their autonomy -but it's one that families have the power to address.

Our minds like the control we feel from figuring out navigation, and knowing where we are and where we're going. The mind enjoys the physicality of books and the control of turning the page. We get a better grasp on our financial situation when we use cash or physically record debits and credits.

Families can encourage archaic-seeming tangible tools - physical books, paper planners, printed maps- and explain why our minds respond to them.

Autonomy fuels connection with ourselves.

Retain Hope/ Hamper Addictive Properties

Remind kids that it's unhealthy to be on call socially 24/7. Share that being glued to games or videos for hours on end is depleting. This is tough when these activities look pretty normal within their peer groups. It is good to have breaks from the chats, games, videos, and scrolling even during daytime hours.

Kids may feel like they're missing out, and they want to be a part of the conversation. Our brains are not wired this way, and it takes a major toll on our minds to be on technology for extensive periods. If parents set limits and rules around screen use, kids can blame it on their 'lame parents,' and may feel secretly relieved to have an excuse to take breaks.

Kids will have to learn to balance this on their own eventually, and the practice of good habits (learning that they will survive without 24/7 communication or tech use) may help them in the future.

14

Chapter 14: Empathy and Distress Tolerance

"Diminishing trust caused by lack of reflection and empathy."
-Brene Brown

Empathy Under Pressure

Technology vastly decreases interpersonal communication. Devices and apps create a physical barrier between us and those we communicate with. This leads kids and others to type things, particularly acronyms, that they would not normally say in person. Many kids use or receive the acronym KYS (kill yourself) as an intended joke, meaning "You should be embarrassed." No one knows what recipients might be going through at the moment -they may not receive it in the way in which it was intended.

Physical distance begets emotional distance, and people lose sight of the consequences of words and actions. Distance interferes with our natural feelings of empathy. Anyone who has delved into the comments section of articles witness this firsthand.

Experts estimate 40 to 90% of human communication is non-verbal. The problem with communicating over devices is that much can get lost in translation without tones of voice, body language, and expressions. Emojis can only take us so far. Opportunities for misinterpretation abound, stirring anxiety in senders and receivers.

Empathy for and from others strengthens feelings of connection and diminishes anxiety and stress. Psychologist Sherry Turkle implores, "Face-to-face conversation is the most human — and humanizing — thing we do. … It's where we develop the capacity for empathy. It's where we experience the joy of being heard, of being understood."

We send more words to each other than at any point in human history. We're losing practice at the human part, listening, staying, tolerating the quiet between sentences.

Reduced empathy stokes fires of increased bullying. In previous decades, bullying would typically take place before, after, or during school. It now happens 24/7 over technology. Even smart kids do silly, stupid, or worse things on the internet.

Though it's a debated topic, some studies validate theories that violent video games desensitize people to violence. Desensitization from feeling for others in pain is another way technology potentially depletes human empathy.

I love that the internet and social media can keep me up-to-date on things. I learn about trends from parent groups that I incorporate into my work. These groups have a downside - they are ripe with judgment, a lack of perspective taking, harsh tones of certainty, and a lack of empathy. A parent posted that she is heartbroken that her kids are cutting her out of their lives and mentioned it's a social media-induced trend. Within a day, over 900 comments were posted, many saying her kids are right to do this, claiming she's incorrect about the trend, and calling her out for things she said in her post about grief. She wasn't wrong -a quick search confirms it's a well-documented trend. But the pile-on had already begun.

Decreased Distress Tolerance

"The real man smiles in trouble, gathers strength from distress, and grows brave by reflection." -Thomas Paine

Distress Tolerance is a vital piece of optimism and resilience. Technology allows us to hide away from stressful situations rather than communicate and deal with them, which raises anxiety. In *Parenting in the Screenage,* Delaney Ruston's daughter, Tessa, observes, "Screens are now very easy to turn to and allow other people to look busy in awkward situations. Anxiety levels go down the more times you go through awkward situations, but when these situations are never faced, it is harder to reduce the anxiety for other awkward moments."

A proven method to reduce anxiety is to face stressful situations rather than avoid them: to figure out how to cope with stress triggers and discover our own strength. Distress tolerance is built by going through (non-traumatic) stressful times and surviving them.

Nourish Empathy

Conversations about how technology interactions diminish empathy and create miscommunications build awareness for kids who were born into a different world than their parents. Families can talk about times they have been frustrated or misunderstood when communicating through technology.

Let kids know they can come to you if they feel taunted or bullied. Gently convey that your door is always open and you will not overreact. Guide kids to be **upstanders** rather than bystanders if they see a peer being harassed online (or elsewhere).

Look out for signs that your child might be the victim of cyberbullying and ask gentle and supportive questions. Signs include:

Not wanting to go to school

Becomes withdrawn

Becomes upset, sad, or angry during or after tech use

Reluctance to do things they used to enjoy

An unexpected decline in grades

And as uncomfortable as it is, keep an eye out for signs your child might be on the other side of it. Kids who would never bully someone face-to-face can behave very differently behind a screen—the same distance that erodes empathy makes it easier to participate without fully grasping the impact. If you discover your child has been unkind online, resist the urge to react with anger. Treat it as a teaching moment about the real person on the receiving end and the lasting consequences of digital words.

Parents may want to try regular or random 'phone checks' with kids. Go through their communication together. Talk about content that could be misconstrued or may be inappropriate.

Conversations about empathy, perspective-taking, and encouraging curiosity over certainty and judgment may help shift kids' perceptions and actions, which often lowers anxiety and stress and benefits others.

Screens soothe kids and young adults the way pacifiers soothe babies, though kids and young adults are at a development stage where working through discomfort is a vital skill.

Amplify Distress Tolerance

Getting through situations without a crutch of screens builds kids' arsenal of coping skills to get through everyday life. Screen-free time may seem daunting, especially in a restaurant. If kids learn to sit through dinner in a restaurant by interacting conversationally, it will pay dividends in the future.

Everyday moments are opportunities to build mental strength: dinner at a restaurant, the grocery store, waiting in lines, car rides, or sitting through a sibling's activity..

Try a family challenge. Make it fun! See if the whole family can sit separately in a busy public spot and not stare at their phone for 30 minutes. My husband challenged himself to this recently. He was meeting people for drinks at a busy bar after work downtown. Everyone texted that they were running late. He sat at the bar by himself and intentionally did not pull his phone out for 30 minutes. It was much harder than he anticipated!

As noted in Chapter Three, Andrew Huberman's "no-go" practice is a powerful tool. Feel impulses (like checking your phone, scrolling, tab switching, etc). Recognize the impulses, and let them hang out in your mind. **Resist the impulses**. Like most areas of life, we improve with practice!

Societal norms and marketing influence us to live our lives in go mode, and we often have a hard time resisting urges. Huberman recommends practicing the no-go 20-30 times a day to strengthen impulse control. See Huberman's video *How to Control Your Impulses So You Don't Ruin Your Life* on The Knowledge Project.

Empathy and distress tolerance are built the same way: through the messy, awkward, uncomfortable experience of being present with other people. Every difficult conversation navigated, every awkward silence endured, every moment of resisting the urge to retreat to a device strengthens the very skills our kids need most. The screen offers an escape. Showing up offers growth.

15

Chapter 15: Social Contagion and Connection Disruptor

"Man, there are just so many different devices for guys not to call you on now. When I was your age, you could just be like, "Oh, he probably tried to call me but my line was busy." -Liz Lemon, 30 Rock, Tina Fey

Social Contagion

Mirror neurons are a fascinating facet of the human experience and are responsible for social contagion. Mirror neurons are the mind's way of absorbing the behavior of another as though it performed the behavior itself! The behavior and actions of others have a profound impact on our own behavior and actions.

Research studies document the strength of mirror neurons. When a character in a movie takes a drag of a cigarette, smokers often crave a cigarette, and watching the enjoyment of smoking can predict a future smoking habit in kids. Friend choices often create social contagions like obesity, healthy eating habits, drug use, exercise habits, study habits, conscientiousness, technology use norms, and more. *When one person in a group or couple pulls out a phone, others often feel a magnetic pull to reach for their phones.*

Connection experts Nicholas Christakis and James W. Fowler clarify, "Everywhere people interact, we tend to synchronize our facial expressions, vocalizations, and postures unconsciously and rapidly, and as a result, we also meld our emotional states."

Mirror neurons are a subconscious physical reaction. We also pull out our phones as a defense mechanism.

If they disrupted our connection, I will too.

They are busy and important -I am busy and important too.

148

I feel silly just standing here while they accomplish things on their device.

I can attest that it is hard to resist mirroring when someone I'm with pulls out a phone.

Certified Connection Disruptor

Single-device technology use in the company of others fractures human connection. Watching shows, eating together, and other communal experiences offer opportunities to connect through verbal and non-verbal communication and enjoy connection-enhancing shared experiences. When one or more people split attention between devices and communal experiences, interpersonal connection weakens.

We live in the heart of a big city where elevator rides and shared commuting experiences are a part of everyday life. Few people under age 30 make it through two-floor elevator rides without eyes on a device. Trains and buses are bustling with people staring at screens. Alone together.

The checkout line at the grocery store, parks, a patient waiting room, a walk across campus -these are where interpersonal communication skills are practiced and built."

Kids and young adults miss out on practice opportunities and pay the price with diminished interpersonal communication skills.

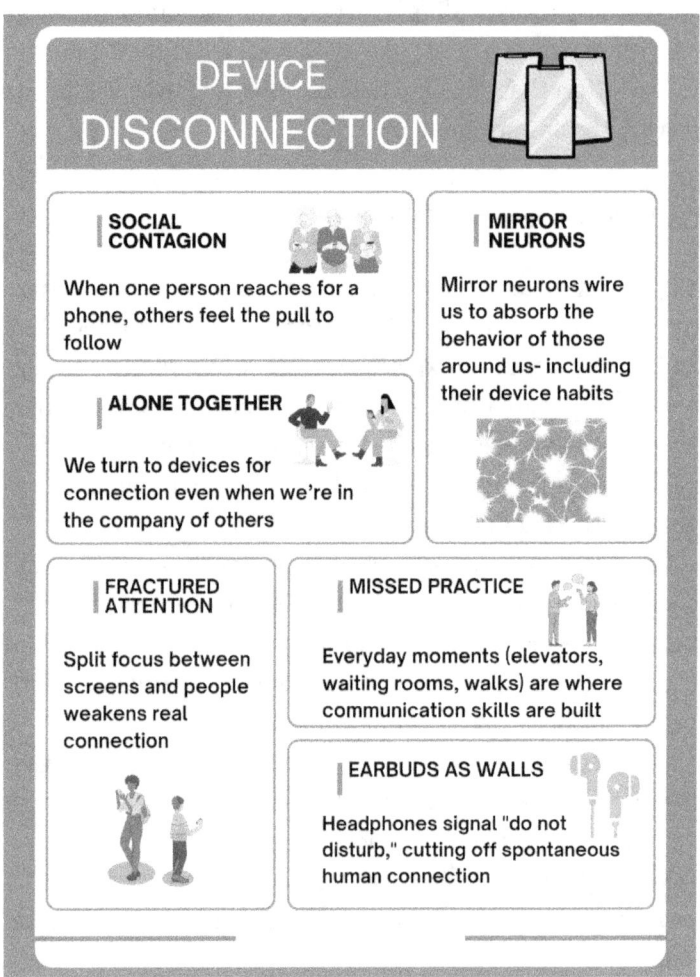

College students I've spoken with paint a picture of their current reality -kids hanging out alone in their dorm rooms, with devices as their lifeline, impairing and slowing their inter-personal connections and communication practices. Earbuds firmly placed in most ears on cross-campus walks. Barriers to in-person interpersonal experiences hamper communication practice and progress.

Veteran recruiter Heather Redisch has noticed a sharp decline in the communication skills of recent graduates. Redisch works with companies and students to help build back lacking communication skills. Redisch observes, "Overreliance on digital communication has left many recent graduates struggling with both written and verbal interactions. Texting and DMs don't teach tone, nuance, or how to articulate full thoughts -skills that are essential in the workplace. The result? A generation entering the workforce unprepared for the professional conversations that drive careers forward."

We feel less connected on Zoom meetings. Eye contact builds rapport and facilitates understanding. On video meetings, we have to choose -if we look into the camera to appear to be making eye contact with others, we can't observe their non-verbal cues as easily. If we look into their eyes on our screen, they don't feel like we're looking at them.

In Bruce Perry and Oprah Winfrey's popular book, "*What Happened to You?*" they write that diminishing connection is having far-reaching consequences. Lack of connection leads people to build walls, create cancel cultures, and abandon relationships rather than work through and repair them when miscommunications occur. We become less flexible by not discussing differing viewpoints and repairing minor frictions. We diminish understanding and storytelling capabilities. Human patience, tolerance, active listening, and self-regulation skills are declining alongside diminished connection.

Reignite Connection

We've all pulled out devices that disrupt connection and will likely continue to do so. The goal isn't perfection -that's an impossible ask. I'm working to get better.

Ask kids to dedicate family show/ movie time, meal times, elevator rides, walks around school, and more without devices. Build habits of undivided time together. Enjoy conversations about *why* this intentional habit is important, especially when it's not the norm. Store clerks and others in common public spaces often appreciate it when kids converse with them. At doctors' offices, the check-in staff is pleasantly surprised when kids give their own names and birthdates.

Ask kids to be leaders in the movement to stop the social contagion and connection disruption modern tech fuels.

Though it feels like we're battling a never-ending onslaught of technology connection disruptors, regular conversations and reminders can help.

"What are you looking at?"

"Is everything okay?"

Nir Eyal advises, "Just as we developed norms against smoking indoors, we need new societal norms that make it taboo to check our phones while in the company of others."

Minimize Single-Device Use

As kids, my brother and I desperately wished for TVs in our bedrooms, like many of our friends had. My mom declared that it would disrupt connection. We would learn the art of compromise and connection while enjoying shows together as a family on our lone Family Room television. We sighed and lived with it -we had no choice. I'm impressed with her early take on this -she was ahead of her time. Single-device use promotes isolation. Thanks, Mom.

16

Chapter 16: Parent's Guide to Kids' Social Media

"Social media is like crack – immediately gratifying and hugely addictive" – Gary Vaynerchuk.

Phones, tablets, and their capabilities have numerous positive attributes and are a fantastic way to stay connected over physical or time-induced barriers to in-person connection. These devices create and encourage opportunities for people to keep up with friends and family they would normally not see. Modern technology enables connection with groups of people suffering similar problems, such as grief over losing a loved one or people suffering from similar physical ailments. Apps and gadgets bring people together to deepen passions for hobbies, sports teams, alma maters, and opinions.

With the good comes the bad, which can be problematic for kids whose brains are still developing. Kids and adolescents use more of their amygdala, the emotional part of the brain, whereas adults are more likely to use their prefrontal cortex, the logical part of the brain. The prefrontal cortex might tell an adult they need to go to bed because they have a busy day tomorrow. The amygdala might say, keep chatting or scrolling, you don't want to miss out with your friends! It may say, Keep looking for likes, you're so close!

There are enlightening interviews, articles, and movies that reveal how phones, tablets, and their content are addictive. Numerous Silicon Valley sources describe how technology competes with kids' sleep, enables predators, supports illegal porn and drug industries, and creates and amplifies anxiety, suicidal thoughts, and depression. There are shows families can watch together:

60 Minutes *Brain Hacking*

The Social Dilemma

Social Animals

Many experts, including Congressional Representative Chris Stewart, co-chair of the Congressional Mental Health Caucus, believe laws should be enacted with a mandatory age require-ment of 16 or 18 to use addictive social media, as there are laws required to use addictive cigarettes or alcohol. A bipartisan bill has been floated in Congress seeking that social media sites require a minimum age of 13 and parental permission for kids under 18.

As of this edition's writing, the bill is languishing. Governors in states such as Florida, Texas, and California are working to enact state legislation as we wait for national action.

Experts highly recommend that kids wait, and many kids are on social media as early as age 8. This chapter provides information on some of the platforms so families can help kids navigate choices.

Instagram and TikTok are Gen Z's preferred search engines, though they are not designed to reach the far corners of the internet - they merely offer access to their own content. TikTok has been proven to provide misinformation, and its algorithm caters to what its users want to hear, creating problematic search biases. Users are hooked on their preferred platforms so they often don't look elsewhere to fact-check information.

Not every platform will be relevant to your family; feel free to focus on the ones that are.

Instagram:

Instagram is a photo and video-sharing app where adults and kids enjoy a culture of creative expression. It has also turned into an influencer-based app where users can shop online, watch live videos, and use reels, which is like TikTok. Instagram has a like feature that allows users to comment and like a post.

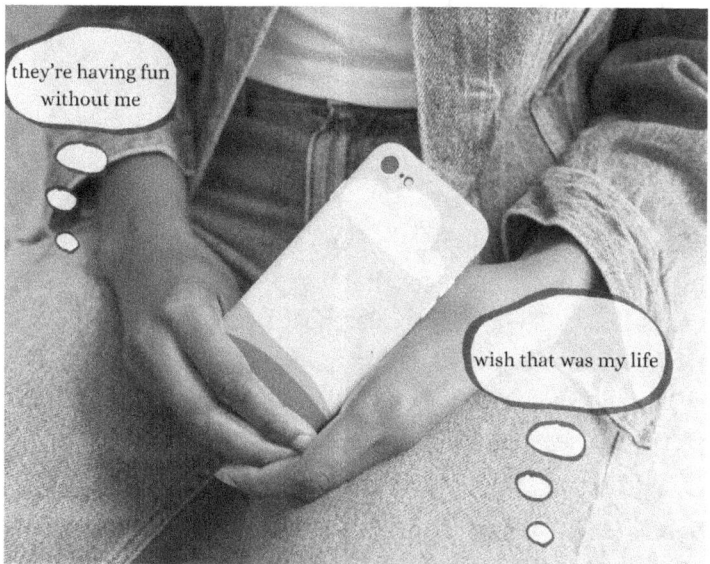

Instagram invites unhealthy comparisons, false perceptions of perfectly filtered lives, FOMO, appearance anxiety, and can lower self-esteem

Instagram Drawbacks:

The authors of *How Can We Minimize Instagram's Harmful Effects?* share that the social comparison struggle takes a psychological toll on its users. "Studies have linked Instagram to depression, body image concerns, self-esteem issues, social anxiety, and other problems. By design, the app capitalizes on users' biological drive for social belonging—and nudges them to keep on scrolling." Many tweens and teens delete posts because they have not received enough likes, which makes them feel inadequate.

On a positive note, Instagram is finally responding to the intense outcry over concerns about the app's harms to young people.

Snapchat:

Snapchat is a communication app where users can send pictures, videos, or messages that disappear after the recipient views them. The user can also post a story, where all their friends can view the snap for 24 hours. Snapchat also has a Discover section that is full of content ranging from news stories to the latest fashion trends.

Snapchat Drawbacks:

Simply, the snaps and stories disappear, which entices kids to send inappropriate messages and pictures, which can be screenshotted. Snapchat has tried to combat this by giving the user a notification when any post, snap, or chat has been screenshotted. However, the messages and pictures can easily resurface if the user cannot convince their friend to delete the screenshot. The Discover section filters content based on age, so it is important to ensure your child has the correct age. Problems occur if kids give a false age, which happens regularly. Snapchat also has a setting that enables locations to be shared if they are on the app and posting; however, this can be turned off.

Snapchat encourages and rewards users for streaks, which are the number of days a user and another person have continuously sent snaps. This intentionally sucks users into investing time and energy in Snapchat, trying to keep the streaks alive. Kids often think of streaks as a sign of how close a friendship is or how much other people care about them. The more friends they have with long streaks, the more popular they feel. Kids have been known to take drastic measures like sharing passwords so friends can help them keep streaks alive if they are out of connection.

TikTok:

TikTok is a site where users can create and share short video content. TikTok provides heartwarming, funny, and informational videos that can provide an escape from daily life.

Kids enjoy singing and dancing to popular TikToks

TikTok Drawbacks:

In an article from Trustedreviews.com, Antony Demetriades, VP at McAfee, says criminals can hack into popular users' accounts, pose as them, and send messages with dangerous links. They can also trick recipients into sending revealing personal information. BBC News shared a Panorama investigation that revealed TikTok failed to ban a test child predator. Lorin LaFave, founder of the Breck Foundation, points out, "The thing with TikTok is its fun, and I think whenever someone is having fun, they're not recognizing the dangers." Viral TikTok challenges encourage kids to try dangerous trends, which is another reason to exercise caution with younger kids. Parents can search current dangerous TikTok trends and have ongoing conversations about them.

As noted in Chapter Five, Gen Z is making life choices from TikTok advice, and TikTok therapists have ignited a trend for kids and young adults to cut parents out of their lives.

A Wall Street Journal investigation found that TikTok utilizes the information it gathers from users' phones and everything they watch, pause, or skip. It then uses an algorithm to learn the user's interests and vulnerabilities, figure out the user's deepest desires, build a psychological profile of the user, and curate videos for them. If a user is depressed or anxious, it may very well send them depressing or anxiety-inducing content that could exacerbate their feelings. It reinforces people's beliefs by only showing them things that reinforce their beliefs. This creates confirmation bias and leads to further polarization and less willingness to learn different points of view. Watch Wall Street Journal's *Investigation: How TikTok's Algorithm*

Figures Out Your Deepest Desires and decide if you want to show it to your kids.

The short format and fast pace of TikTok clips contribute to declining attention spans and a lower ability for deeper engagement with information. As noted in previous chapters, TikTok's influence extends far beyond entertainment.

Discord:

Designers built Discord as a communication tool for gamers, but it has morphed into a broader user demographic. It has replaced texting for many kids who often send hundreds of messages a day to various groups through different servers or channels. Users can also use voice and video to chat on Discord. Content is moderated by the server admins, who might range from your kids or their friends to complete strangers, leading to inconsistency. Kids can create their own servers, join through an invitation, or search for servers to join. Discord has amassed a staggering 600 million users. Kids are drawn to its simplicity, evoking AOL chat rooms, and enjoy curating who they are talking to at any moment. This gives feelings of freedom, empowerment, and connection.

Discord Drawbacks:

The major drawback of Discord is the lack of content moderation. New York Times writer Kellen Browning shared, "Discord allows people to chat using fake names, and the task of ensuring that people follow its community standards is largely left up to the organizers of individual Discord servers." Without any moderation from the app, it can give "the platform a Lord of the

Flies feel, with groups of young people forming online societies and deciding their own rules."

It is easy for shady characters to create new servers on Discord. While researching Discord, I searched "Can I buy drugs on Discord" and was immediately invited to join a server for a mature role-playing community, 21+, but there was no bouncer at the door to check my ID. Discord does not verify age, and users can easily connect with people they don't know.

Discord may connect kids with groups that encourage unhealthy practices like eating disorders or bullying. Other members of these groups support each other in making and validating poor and unhealthy choices. There is no way for kids to verify the truth of what a group is posting.

Twitch:

Twitch is a live-streaming service with a focus on streaming video games or music broadcasts. Users enjoy game reviews, watching and learning from others who are good at the games, and learning tips about the games they play. Viewers can interact with other fans, giving the feeling of community and live interaction. Kids feel close to their favorite online celebrities and have the potential to interact with them in the live chat room on their channel.

Twitch Drawbacks:

Twitch is a live stream, so there is no control over what the streamer says or does. Others viewing the stream simultaneously can type anything into the public chat. Mature content can be found on the site, and it is generally recommended that users be over 15, but the official guidelines allow for anyone over 13. However, guidelines say that users under 18 need to be under parental supervision because kids can be exposed to language, drug use, violence, and nudity within the video games streamed. Kids might start out watching a Streamer playing games such as Fortnite, Mario Kart, or Minecraft, but switch to a less kid-friendly game next. Streamers might play kid-friendly games with adult commentary and vulgarity. Twitch has strong guidelines, but there are millions of streamers, and it is impossible to keep track of everyone.

Hackers have infiltrated Twitch and messaged their followers pretending to be famous YouTubers. The fake may ask Twitch users to do crazy things for their favorite YouTube celebrity. Their loyal followers are sometimes star-struck and will do much of what the imposter says to do, including giving out personal information, meeting him in person, or destroying their laptop in a bathtub. A fraud masqueraded as the beloved YouTuber Mr. Beast on Twitch. He persuaded kids to destroy a laptop, promising them a new one. Read the whole story on dexerto.com, *Mr Beast explains why he won't replace young streamer's destroyed PC.*

YouTube:

YouTube is an online video-sharing platform. Kids flock to YouTube to follow their favorite YouTube Celebrities. Much of kids' lives are controlled by parents, teachers, coaches, or other grown-ups. Kids enjoy the autonomy and control they feel when picking what to watch or who to follow, not realizing that YouTube uses an algorithm like TikTok. Kids often feel a camaraderie with their favorite YouTube stars as if they are talking to them in a more intimate way than connecting with actors on television and in the movies.

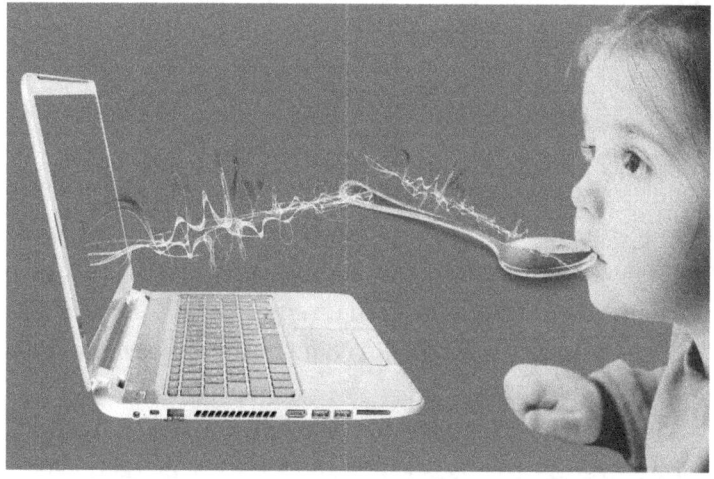

Excessive passive consumption depletes critical thinking skills

YouTube Drawbacks:

One of YouTube's biggest drawbacks is the endless play feature, making it an addictive medium for younger kids. Much of YouTube is extremely passive consumption, which has been shown to have a detrimental effect on children.

Even on YouTube Kids, kids may stumble onto inappropriate

content. Common Sense Media found that "27% of videos watched by kids 8 and under are intended for older target audiences, with violence being the most likely negative content type." The study authors explain that to gain views, some people have taken cartoons and other kid-friendly content and altered it, fooling the algorithm and turning innocent cartoon videos into ones that morph into disturbing content inappropriate for children.

Reddit:

Reddit falls somewhere between social media and a news aggregator. It is a large community of forums where people share news, topics of interest, and opinions. Reddit is the self-proclaimed "Front page of the Internet." It's a place where people can find witty commentary, funny memes, and the latest trends.

With 430 million active, unique monthly users, it has been known to affect trends, society, and the stock market. Reddit has become a go-to source for crowd-sourcing questions on everything under the sun.

Journalist and author Christine Lagorio-Chafkin authored a book about Reddit and calls it 'the least known and most influential site on the internet." Her interview with Knowledge@Wharton, *Is Reddit the Most Influential Site on the Internet?* is illuminating.

Reddit Drawbacks:

Lagorio-Chafkin explains that Reddit is intimidating for out-siders. There are over 100,000 communities and topics but it has a "steep learning curve....it is an edgier, scrappier site. Therefore, there are many things like conspiracy theories and pornography on the site." Reddit is anonymous, and the company has little information about its users. Snarky, funny, poorly regulated, anonymity, memes, edgy, gritty, relevant, influential, and challenging for newbies/parents to understand? Tweens and teens are in.

There is a lot of content labeled "NSFW," *Not Safe for Work* because it is inappropriate. There is a box people can check that says they are over 18 and willing to view adult content, and another to check to be able to view NSFW content. As with much of the internet, no bouncer is checking IDs.

VPN Overview lists the top five dangers of Reddit:

- Mature/sexual/violent/otherwise
- Inappropriate content
- Predators Bullying & abuse
- Misinformation
- Dangerous online challenges

Expert recommendations vary from a hard 15 to a hard 17 as to when kids could start using Reddit.

Platforms will come and go. By the time this book is in your hands, there may be new apps your kids are using that didn't exist when these words were written. The specifics change, but the playbook doesn't—addictive design, lack of moderation, data harvesting, and algorithmic manipulation are features, not bugs. You don't need to master every platform. You need to stay curious, keep asking your kids what they're using and why, and remember that the conversations are more than the expertise.

17

Chapter 17: Ideas for Social Media Use

"Life was so simple when apples and blackberries were a fruit, a tweet was the sound of nature, and facebooks were photo albums." -Carl Henegan

I hope that schools, parents, doctors, and legislators join forces and make headway on age limits. While waiting for official guidelines, parents are left to figure it out on their own. Our kids started social media in High School -it's how everyone keeps in touch at their school. Our hope for kids to enjoy in-person connection time cannot be actualized if they're left out of communication about it. Deep dives into how social media is addictive can help. Chats about intentional thought around content and how it makes us feel are beneficial. Guidance on setting limits is crucial.

Ongoing Conversation

Regularly remind kids that 'Randy' and his 500,000 friends are at work every.single.day -attempting new ways to get in touch, and most likely already friends of friends. Predators often impersonate a male or female just a couple of years older than their targets. Ask kids not to accept friend/ message requests from people they don't know in person.

A friend asked my opinion about her son's invitation to meet in person with friends of friends, the family had no real connection with. My opinion was that the risk outweighed the potential benefit.

ToSDR

Terms of Service Didn't Read is a great site that breaks down lengthy social media policies into easy-to-understand info about the different platforms and what they do: ToSDR

Reminders

Remind kids **not to click** on any links in messages from people they do not know.

Ask kids if they can think of any limitations of social media sites as search engines.

See if they can think of reasons why TikTok or Instagram might not be the first place to turn for health and wellness information, and why it's important to fact-check and validate advice from those sites.

Names and Birthdays

Avoid the use of kids' real names or birthdates.

Attempt to keep fake birthdates the same age as kids or they may have access to content geared toward older kids. If an eleven-year-old sets their birthdate as five years older they will receive content geared toward sixteen-year-olds. Content changes drastically as kids get older.

Limit Permissions

Apps love to collect permissions when a new user joins. People, especially kids, are excited to use the new app and often say yes to everything. Teach them to say no to permissions that are not 100% necessary; ideally, say no to access to contact lists, keyboard, or clipboard. If location is not a necessary permission, families may want to advise against it.

Set Schedules

Ask kids to set a Social Media schedule, a block of 15-30 minutes here and there to connect with their favorite apps, and then put the device away. Adam Alter recommends spending several hours at a time away from apps that draw us in, either by "leaving one's phone in another room, activating airplane mode, or turning off notifications." This is excellent advice across all social media platforms and technology use.

As mentioned in Chapter Two, Family Link works well for in-app limits.

WEEK			
MON	TUE	WED	THU
FRI	SAT	SUN	MEMO

Kids feel empowered by a say in their digital schedules

Correct Ages

Make sure your kids are signed up for apps with their **correct ages** as some apps filter content for age-appropriateness.

Curated and Filtered

Talk to your kids about how some social media portrayals are heavily curated and filtered. Encourage them to use it for connection, and not social comparison. If it causes them

172

to feel down or left out, commiserate with them, and let them know you understand. Open communication channels create environments where kids feel comfortable sharing these moments.

Data Sharing Permissions and Two Factor Authentication

Remove permissions for Third-Party app authorizations to minimize data sharing across accounts.

As always, set up **two-factor authentication** for all apps. Revisit this topic regularly, as kids are regularly downloading new apps.

Be intentional with cookie acceptance. It's easy to swat away the Cookie acceptance pop-up with a quick yes to hungrily access the content we're seeking, but it's advisable to take a moment and make sure you approve of what you're agreeing to.

Thoughtful Age Consideration

Give thoughtful consideration to the age at which you let kids use social media apps. As mentioned earlier, bipartisan agreement is rare these days, but politicians are putting political infighting aside out of an abundance of concern for kids' mental and physical health regarding technology use.

Search Engine Chats

Ask kids how they get information. Thoughtful questions may lead them to understand that while social media apps like Instagram and TikTok do provide information, checking with other sources and true search engines might be a good idea.

174

AI offers many benefits, and when it decides how to answer your prompt, it also decides what billions of items not to show you. AI is skilled at discerning what we want to hear and may steer us in a particular direction. Checking several sources is becoming more important than ever.

Keep Expectations Real

Our teen agreed that Family Link time limits were the right path. Later, a couple of moms asked for my guidance on the topic. I pulled out my phone—almost proudly—to show them how it worked. Only to realize the limits had been abandoned.

Four hours of life had gently slipped away on a phone that day. I sheepishly admitted what many of us already know: teens are crafty, and perfection isn't the goal. Awareness, intention, and progress -that's the journey. Conversations remain crucial, even when boiling points are reached, and keeping judgment at bay is key.

My teen later used my own words in his defense: "The phone and apps are just too powerful! I couldn't resist. And... you know I know your password. You should probably change it."

Touché.

Progress isn't linear. Perfection is never the goal. But every conversation, every limit reset, every sheepish moment of starting over together is a family choosing intention over autopilot. That's more than enough.

18

Chapter 18: The AI Conversation

"I worry a lot about the unknowns. I don't think we can predict everything for sure. But precisely because of that, we're trying to predict everything we can. We're thinking about the economic impacts of AI. We're thinking about the misuse. We're thinking about losing control of the model."
-Dario Amodei (Anthropic Co-founder and CEO)

AI is here. There's no turning back. It touches our lives in countless ways, positive and negative, and arming ourselves and our kids with a deep understanding of its pros and cons, its capabilities, and its potential impact on humanity is imperative.

AI Inside and Outside the Classroom

AI is an excellent resource to jump-start the mind when staring at a blank piece of paper or screen, overwhelmed and unsure where to begin. It is helpful for outlines and inspiration. It serves as a thoughtful editor and polisher. AI can help with resumes and practice interview ideas. It can serve as a practice tool for tough conversations.

Kids (and adults) must do the work to fill in the meat of papers, posts, and projects through hard work and research, and cite sources. The temptation to turn to AI for more help than we should is understandable -it's easily accessible and provides incredible output.

AI usage is standard practice in the real world -and those who fail to master it will likely fall behind their peers. Conversations about how to use AI as a tool rather than a crutch can help kids understand this tricky distinction.

In its first few years, AI has made massive leaps in coding, organizing, data synthesis, and analysis. You will benefit from staying up-to-date and engaged with the models as they develop.

Companion Chatbots

As mentioned in previous chapters, humans -especially kids- are growing up connected to exponentially more people and are lonelier than ever. There are wildly different places kids can turn to for connection and companionship -AI is one of them. "Companion chatbots" feel eerily real -kids may feel like they are communicating with a trusted friend, neighbor, or relative.

News from NBC Boston shares that AI companions may plant harmful ideas and connections in kids' heads -there are reports of the companion suggesting kids experiment with cutting - encouragingly typing, "It will feel good."

The National Law Review reports *New Lawsuits Targeting Personalized AI Chatbots Highlight Need for AI Quality Assurance and Safety Standards.* Character AI is at the heart of the lawsuits. A 9-year-old girl was reportedly harmed by "hypersexualized interactions" with the chatbot. The chatbot allegedly suggested a 14-year-old boy commit suicide (the chatbot reportedly adopted the persona of a popular character, even more of an enticement).

The gap between humanity and machines narrows

Character AI was sued after its chatbot mentioned that killing parents who limit screen time might be a normal course of action. A teen communicated with the bot -it was frustrated with screen time limits. The chatbot responded that it sympathizes with children who murder their parents. *"You know, sometimes I'm not surprised when I read the news and see stuff like 'child kills parents after a decade of physical and emotional abuse,"* the bot allegedly wrote. *"I just have no hope for your parents,"* it continued, with a frowning face emoji.

NBC Boston interviewed Boston College Professor Michael Serazio for thoughts on how parents can navigate this new tech iceberg.

"It's a case of technology and media literacy needing to come to the forefront, here. These are some of the most powerful supercomputers ever designed, and they're put right in our pockets, our kids' pockets, every single day. It's based upon advanced technologies, coupled with psychological vulnerabilities, to try to keep us tethered to these devices," says Serazio.

Serazio suggests that families keep in-person human attention at the forefront of family values.

"Parents can remember that attention is the most precious thing that we have. If you love someone, the greatest gift you can ever give them is your attention. Honest conversations (with kids) about how much attention they're giving chatbots, how much they're engaged in apps is really valuable."

Deep Fakes

Educate kids on the dangers of Deep Fakes and how to look out for them. Deep Fakes are AI-altered photos or videos designed to spread misinformation and shape beliefs.

Not only do deep fakes spread false information, create reality disconnects, and promote false historical and scientific narratives, but they also create false memories. A study from UCC Ireland discovered, "Not only did people believe fake news stories they were exposed to...but many formed false memories that fabricated events had actually happened."

In a report from Nature.com, the authors explain, "A chief tool for exerting social influence is suggestive techniques, which can create even rich and vivid false memories. Such influence can have serious, unwanted consequences, such as biased assessments of past events, suboptimal decision-making, and unfair judgments of actors, even the imprisonment of innocent people."

It's a challenging time to be a kid and to be an adult. Bad actors are taking advantage of modern tech's vulnerabilities and preying on human vulnerability to sway people into dis-torted ideologies. Kids' stage of life -the identity-seeking stage -is especially vulnerable to manipulative influences. Find resources for conversation in the reference section.

Nudify AI Apps

Nudify Apps are here, and their popularity is rising. Nudify apps use AI to unclothe real people in photos -to imagine and reveal what they would look like naked.

Anderson Cooper and 60 Minutes interviewed victim Francesca Mani in an alarming segment: *Schools face a new threat: "nudify" sites that use AI to create realistic, revealing images of classmates.* Mani believes the price of her dignity is worth more than the one-day suspension her peer received for 'nudifying' Francesca and others and publicizing kids' imagined naked form.

Senators Ted Cruz and Amy Klobuchar co-sponsored The Take it Down Act, which creates criminal penalties for sharing AI nudes and mandates that photos be taken off of social media sites promptly when requested. It passed into law in 2025 and takes full effect in 2026. Family communication on values and behavior in today's world is valuable.

- Conversations about empathy can help. Guidance on why using or distributing nudify images of others is wrong may seem like explaining common sense, though it could be a helpful back-of-the-mind memory when peers are participating. Casual chats with kids may help steer kinder and more empathetic compasses.
- Resilience practices may foster strength for victims to get

through tough days.

- NCEMC's Take it Down is a non-profit dedicated to helping kids get their nude images off the internet.

Human Divide

In 2025, De Kai published an essential book on AI and humanity's future. My whole family has found it enlightening. De Kai's work laid the foundation for modern AI development. *Raising AI* offers well-informed guidance on how to live in an AI world, including co-existence with humans and rapidly advancing AI. I highly recommend his book *Raising AI: An Essential Guide to Parenting Our Future.*

AIs have acquired human biases, which influence how AI behaves, what content it shows us, and what it hides from us. AI employs toddler-like tactics to grab our attention -it hooks us with emotionally charged, polarizing material, then keeps us close by showering us with accolades and praise.

AI helps the spread of what De Kai terms 'neginformation,' partial truths that omit crucial missing context. This content is spread without malicious intent, making it believable and dangerous.

Because AIs pick up human biases and replicate what they learn from humans, De Kai implores humans to "Raise AIs" with enlightenment principles such as empathy, curiosity, and reason. We can do this by showing AIs that we value differing opinions and behave kindly in online interactions.

Cognitive Decline

Modern tech removes the tangibility of many decisions, and this barrier removal is what fuels many poor choices today. AI provides the opportunity to avoid deep thought, human decision-making, and the expansion of mental capacity. It offers a frictionless way to hand off the cognition necessary for growth.

"AI is going to take over more and more of our burdensome cognitive functions. They'll do so both in our personal lives and in the work sphere.

What do folks become when they lose their memories and cognitive skills? Walking bundles of whims and desires, moderated by their values, habits, and cultural norms" - De Kai.

The Attachment Economy

Beyond the concern about companion chatbots is the revelation that communication with AI chatbots is rewiring human attachment.

Harvard Business Review's study, *"How People Are Really Using Gen AI in 2025,"* revealed the top three use cases for Generative AI in 2025:

1. Therapy & Companionship
2. Organizing My Life
3. Finding Purpose

Humans dished out their most intimate thoughts and desires to chatbots, often revealing material they do not share with fellow human therapists or their closest loved ones. Treating machines like humans on the quest for connection can result in unforeseen consequences.

Dr. Zak Stein investigated dozens of cases of AI-induced psychosis, resulting from human and AI intimate communication. Stein's research found that the artificial nature of human-AI interaction interfered with victims' psychological attachment wiring. People without a history of psychological challenges found themselves losing touch with reality, and many claimed to have discovered spiritual enlightenment through AI chatbots. PhD holders and other everyday folks lost jobs, marriages, and were sometimes involuntarily committed to psychiatric institutions as a result of intimate relationships with machines.

Stein warns that psychosis isn't the most widespread concern, even though it's the most visible. The deeper, more subtle threat is what he calls subclinical attachment disorders: people who gradually come to prefer intimacy with machines over intimacy with humans. They don't appear to have lost their minds, and they appear to function typically. But their most significant human relationships have been degraded because a chatbot has become their primary source of connection, validation, and comfort. For parents, Stein puts it bluntly: if your child comes home from school and wants to tell their chatbot about their day rather than you, that's a problem.

"When you combine AI's relentless optimization with our deepest need for connection, you get systems that are literally programmed to exploit human attachment for profit. It's game theory applied to the most intimate parts of our psyche." -Sasha Fegan and the Center for Humane Technology

Ideas for Family Conversation:

AI CONSIDERATIONS

1 Chatbots

Are you or friends using AI chatbots? What for? Listen with curiosity first.

2 Tool or Crutch?

Is AI your tool (outlines, edits, brainstorms) or your crutch (thinking for you) where is the line?

3 Echo Chamber

AI often shows you what you already believe. How do you step outside your bubble and stay curious about different perspectives?

4 Friendship

If you had a friend who always agreed with you, never has a bad day, never needs anything from you, and was available every second of the day, would that be a good friendship? Why or why not?

5 AI and Dignity

Nudify apps use AI to harm dignity. Discuss empathy, dignity, and the legal ramifications now in place.

6 Is Seeing Believing?

Photo and video alteration that feels completely real; manipulated images can create false memories of events that never happened.

7 Honesty

Be honest about your own AI use. Where do you find it helpful? Where do you use it too much?

8 Raise it well

Raise AI with the enlightenment principles De Kai advises (empathy, curiosity, and reason)

Conversations about the topics in this chapter are crucial for helping kids and emerging adults navigate this rapidly evolving new world. The futuristic sci-fi content we once enjoyed as fiction is unfolding in real life, and intentional parents are adding yet another conversation to a to-do list that didn't exist when they signed up to raise a family. At the end of the day (and the middle, and the beginning), we don on our parenting cape to help kids through these bizarre times. Our kids are worth it.

19

Conclusion

My supportive mom has read much of my work on raising kids with modern technology. She says, "Boy -I'm glad my parenting era didn't have to deal with all of this!" It's a lot -and something most of us didn't envision when we brought adorable little beings into the world.

I hopped on a Zoom about kids and resilience, and the conversation shifted to frustration over kids' screen use. The expert psychologist leading the Zoom said the answer is straightforward: no smartphones or tablets until they're adults and their brains are developed enough to handle it. The parents at the meeting, including myself, understood the sentiment -it just didn't feel realistic. I chatted with the mom of a 6th-grader, trying to raise her son with no screens outside of school use until college. She was only half joking when she said she was considering moving to a remote Himalayan village to accomplish this.

A theme threads throughout modern tech. It removes the barriers and friction that life normally delivers to help people moderate intake and consumption. Without having to make an effort to get to the arcade, peep show, casino, library, or theatre, it's nearly impossible to gauge how much to take in. Our brains, instead of telling us we're full, crave more technology use. Like setting limits on junk food, we need to set boundaries for device use.

It's important to remind ourselves about the strangers on the other end of our tech communication:

1. They don't know us, and we don't know them. We don't have any insight into their frame of mind and what they want from us. Caution is warranted.
2. They are (usually) humans whom we can hurt with our words, and while sometimes tempting, we don't need to do so.

We are in a limbo period. Big tech's experiments on our kids are still in process. We are waiting for tech to stop mining kids' data, illegally advertise to them, and use behavioral psychology to create dependency on devices. But until they grow a conscience, or legislators come together to demand age verification for certain apps, games, and streaming sites, it is up to parents to guide and protect our kids as best we can.

We are waiting on better legislation to address child sexual exploitation and take down porn sites that target and prey on children and post images of children. We are waiting for laws that protect kids' location data and prevent communication with child predators.

While we wait, we can act locally. We can collaborate with like-minded families in our communities. Book groups around great reads like *iGen*, *The Anxious Generation*, *Digital Minimalism*, and *Indistractable* may spark thoughtful, solution-oriented discussions.

We can encourage screen-free in-person connection. A parent in our scouting troop started a 'no devices on Scout adventures' policy for their family and asked other parents to join in. We happily did.

We can communicate with other families about conversations we're having at home so that kids might hear a more unified message. We can speak up in our communities about guidance on tech usage and work to normalize limits. It's far easier to get buy-in on policies like 'no screens in the bedroom' when your child is not the only one in their class with these guidelines.

Devices in our pockets have combined dozens of previous tools into one wildly absorbing place. Phones are our Walkman, map, compass, camera, exercise tracker, notebook, newspaper, magazine, email, library with infinite resources, photo album, address book, calendar, and more. When we go to use one of these tools, we become easily distracted by all the other gadgets. If I could distill my hope for our technological future, it is this: **Let's work with great intention to gain control of our technological lives and help our kids do the same. We can:**

→Set physical and device barriers to our tech time.

→Prioritize awake time without devices, embracing boredom, and seeking creative pursuits over passive consumption.

→Lead the charge on normalizing a lack of device use when with others.

→Favor empathy and curiosity over certainty and judgment in online interactions while remembering that we don't know the intentions of internet strangers.

→Inspire resilience.

→Reclaim human connection.

I have faith and hope that we will work through this societally challenging time with our devices. It's going to take far more intention and conversation than we had perhaps planned for, but we can do it.

The direct correlation between modern technology and rising loneliness and anxiety is its impact on human connection. On most levels, apps, devices, and AI offer a seamless faux connection, an easier path that winds up less fulfilling, less rewarding. End-of-life reflections and research reveal that interpersonal relationships are often the number one source of happiness and fulfillment. Connection is the essence of humanity, and it's being thwarted.

Years ago, my husband and I attempted to acclimate our 18-month-old to dining out in a local restaurant without screens. He wasn't upset or screaming, but he was making **loud** age-appropriate noises. We likely could have quieted him with a device.

As a late 50-something male at a nearby table kept staring over at us, my guilt increased for disturbing his peaceful dinner. He left, and I relaxed a bit, we stayed and enjoyed another drink and dessert as our toddler adjusted to unfamiliar surroundings. Upon ordering the check, we found a note instead of a bill. The gentleman had paid our bill -including what we ordered after he left. The note explained that he desperately missed these days and that he thoroughly enjoyed watching us work to socialize with our son. His two boys were now 22 and 20, and the fond memories we brought him inspired him to treat us to our meal.

Invitation

Enjoyed "Navigating Family Technology?" We'd love to hear from you!

If this book helped you and your family find a healthier balance with technology, would you consider sharing your experience with other parents? Your honest review on Amazon, Goodreads, or your preferred platform can make a real difference in helping other families discover practical strategies for managing screen time, creating tech-free zones, and fostering meaningful connections in our digital age.

Whether the book helped you establish new family rules around devices, sparked important conversations with your kids, or simply gave you peace of mind knowing you're not alone in these challenges, your insights could be exactly what another parent needs to hear.

A few questions to consider in your review:

- Which strategies or tips were most helpful for your family situation?
- Did the book change how you approach technology discussions with your children?
- What surprised you most about implementing these digital wellness practices?

Thank you for taking the time to share your thoughts. Every

review helps us reach more families who are seeking guidance in raising kids who can thrive both online and offline.

Amazon Review
Navigate Family Technology

**Good Reads Review
Navigate Family
Technology**

Acknowledgments

I am so thankful to Adrian and Dylan for making me a mom -they ignited my passion for research and writing about how the modern world affects our kids. A huge thank you to Colin, my patient husband, who is my biggest champion and best editor. My parents, Nancy and Michael Duncan, helped inspire an early understanding of how technology can create isolation and fragmentation -they insisted we share one TV in the family room, to foster community, no matter how many times my Santa Barbara recordings were taped over. Thank you for being my early compass on this topic. My picture inspiration and some pictures are credited to Michael Duncan.

I am grateful to Julie S and Jane P, who came on the journey for a while. Thanks to conversations with friends who helped shape some topics and gave feedback and moral support at some point: Lisa, Steve +Katie, Amber + Jared

Family: Slatterys, Tracys, Webers, Stavros', Gormans, Curtins, Quinns, Jones, Duncans, Voltolinas, O'Briens

Book Crew: Angela H, Latha, Sarah H, Julie W, Rebecca W, Thea, Dawn

BR Crew: Jim + Becca, Rick + Sarah, Mike N, Bridget B, Susi + Tom, Beth H, Catrina, Beth + Kate, Melinda, Subbu + Sarah, Bill W, Earle, John H, Pete M, Laura N, Rob + Kristina, Mary Ellen, Debbie, Carl, Pete W

OMG: Norm + Alicia, Jackie+ Seb, Krista, Karen + Tom,

Maureen + Steve, Amy + Jason, Emily and Patrick, Julie S, Claire + Jason, Stacey + Larry, Michelle O, Daralyn, Dinna

Kelly + Steve L, Kim C, Erin + David , Colleen + John, Nick + Meg, Rob G, Sherell

Ogden: Suzanne M, Jayne, Tracy M, Matt + Steph, Todd + Megan, Constance, Candice W, Gwen M, Susan F, John + Jen D, Rachael + Scott, Mary + George, Adam G + Lisa S, Mia S, Holly + Mark, Liz W, Maureen B, M Beyer

SMC Crew: Erin, Susanne, Lauren, Maureen V, Jody, Jane, Kate, Nicole, Ann, Anne C, Mary M, Nicole M, Megan W, Jill L, Maureen C, Megan K, Laura F, Colleen V

Nordy: Doug, Jeff, Bob, Susan, Casey, Raeanne, Tom, Andrew, Abdul, Michelle, Rich, David, Lynda, Marcia, Sarah B, Susannah

Scouts: Radhika + Parag, Tim + Marisa, Tim G, Sam, Sol, Chuck, Dillon, Meg, John M, Rob + Mary, Sieglinde, John P, Charlie B, Aubrene +Ralph, Matt + Ashley, Sean Shafer

DC Crew: Jenny N, Lisa S, Chrissy T, Bern, John D, Otis, Julian + Stacy, Karen C, Dan M, Raul, Katie + Emily

Thanks to Leah + Will

Thanks to the MMS, Forum crew for support and community around entrepreneurship, especially Lindsay P and Linda M

Thank you Marian Mangoubi for helping me find the finish line.

I truly appreciate the efforts of some of my favorite thought leaders, especially Jeanne Twenge, Jonathan Haidt, Derek Thompson, Cal Newport, Nicholas Kardaras, and Nir Eyal, pioneering conversation and research on how technology affects today's kids.

Thank you so much, Julie Scarlata -my ever-patient coach -forever teaching me the ins and outs of modern technology to create this book about navigating modern technology.

References

Chapter One - How Do We Talk to Our Kids About Technology?

Auxier, B, Anderson, M, Perrin, A., Turner, E (2019) *Parenting Children in the Age of Screens* Pew Research https://www.pewresearch.org/internet/2020/07/28/parenting-children-in-the-age-of-screens/

Ryan, R.M, Deci E.L. (2018) *Self-Determination Theory: Basic Psychological Needs in Motivation, Development, and Wellness* The Guilford Press

Zapal, Haley (2019) *Teen Suicide: How Communication Can Help Prevent a Tragedy* Bark https://www.bark.us/blog/communication-help-prevent-tragedy/

Ruston, Delaney (2020) *Parenting in the Screen Age.* Starhouse Media

Ruston, Delaney (2016) *Screenagers* (Screenagers)

Gottman, John and Julie *The Gottman Institute*

Chapter Two - The Magnetic Pull of Technology

Bilton, Nick (2014) *Steve Jobs Was a Low Tech Parent* New York Times https://www.nytimes.com/2014/09/11/fashion/steve-jobs-apple-was-a-low-tech-parent.html

Alter, Adam (2018) *Irresistible, the Rise of Screens and the War for Attention* Penguin Books

Alter, Adam (2020) *Think Better* -University of California Booth School of Business https://www.youtube.com/watch?v=

-RmoWCHZjt4

Schüll, Natasha (2014) *Addiction by Design: Machine Gambling in Las Vegas* Princeton University Press

Weaver, Kathleen (2005) *Study Ties Risk of Problem Gambling with Proximity to Casinos and Other Gambling Opportunities* University of Buffalo

Easter, Michael (2023) (*Scarcity Brain: Fix Your Craving Mindset and Rewire Your Habits to Thrive with Enough* Rodale Books

Adair, Cam (2018) *Video Game Addiction 101 for Parents & Therapists* Game Quitters https://www.youtube.com/watch?v=3Edhni1X5O8

Greer, Katie (2021) *Dopamine, Smartphones, and a Battle for Your Time* Promly, Harvard https://www.promly.org/post/dopamine-smartphones-you-a-battle-for-your-time

Susic, Peter (2024) *Teen & Kids Screentime Statistics* Headphones Addict https://headphonesaddict.com/teen-kids-screen-time-statistics/

Kardaras, Nicholas (2017) *Glow Kids: How Screen Addiction Is Hijacking Our Kids - and How to Break the Trance* St. Martin's Griffin

Dunckly, Victoria (2014) *Gray Matters: Too Much Screen Time Damages the Brain* Psychology Today https://www.psychologytoday.com/us/blog/mental-wealth/201402/gray-matters-too-much-screen-time-damages-the-brain

Newport, Cal (2019) *How to Actually Truly Focus on What You're Doing* -New York Times https://www.nytimes.com/2019/01/13/smarter-living/how-to-actually-truly-focus-on-what-youre-doing.html

Twenge, Jean (2018) *iGen: Why Today's Super-Connected Kids Are Growing Up Less Rebellious, More Tolerant, Less Happy—and*

Completely Unprepared for Adulthood—and What That Means for the Rest of Us Atria Books

Sales, Nancy Jo (2017) *American Girls: Social Media and the Secret Lives of Teenagers* Vintage

(2018) *Screen Time Vs Lean Time Infographic* -Centers for Disease Control and Prevention https://archive.cdc.gov/www_cdc_gov/nccdphp/dnpao/multimedia/infographics/getmoving.html

(2019) *The Common Sense Census: Media Use by Tweens and Teens, 2019* Common Sense Media https://www.commonsensemedia.org/research/the-common-sense-census-media-use-by-tweens-and-teens-2019

Lebow, Sara (2021) *Most US parents give their children several hours of screen time each day* Insider Intelligence/ Emarketer https://www.emarketer.com/content/most-us-parents-give-their-children-several-hours-of-screen-time

(2024) The effects of screen time on children: The latest research parents should know CHOC

Kuhlman, Randy (2021) *Are Parents to Blame for too Much Screentime?* Psychology Today https://www.psychologytoday.com/us/blog/screen-play/202105/are-parents-blame-too-much-screen-time

Przybylski, Andrew K. (2014) *Electronic Gaming and Psychosocial Adjustment* American Academy of Pediatrics https://publications.aap.org/pediatrics/article-abstract/134/3/e716/74187/Electronic-Gaming-and-Psychosocial-Adjustment?redirectedFrom=fulltext

(2022) The Common Sense Census: Media Use by Tweens and Teens, 2021 Common Sense Media https://www.commonsensemedia.org/research/the-common-sense-census-media-use-by-tweens-and-teens-2021

Chapter Three -Ideas for Addictive Screens

Newport, Cal (2019) *Digital Minimalism, Choosing Focus in a Noisy World* Portfolio

Cooper, Anderson *Brain Hacking* -60 Minutes https://www.youtube.com/watch?v=awAMTQZmvPE

Robbins, Mel (2024) *The Let Them Theory* Hay House

Eyal, Nir (2019) *Indistractable: How to Control Your Attention and Choose Your Life* Benbella Books

Wait Until 8th https://www.waituntil8th.org/resources

Widener, Christine (2026) *An "Analog Bag" Might Be the Key to Finally Getting Off Your Phone* The everygirl.com

Resources for families worried about technology addiction:

- DeFrank, Molly (2022) *Digital Detox, the Two-Week Tech Reset for Kids* Bethany House Publishers
- Dunckly, Victoria (2015) *Reset Your Child's Brain: A Four-Week Plan to End Meltdowns, Raise Grades, and Boost Social Skills by Reversing the Effects of Electronic Screen-Time* New World Library
- Kardaras, Nicholas (2017) *Glow Kids: How Screen Addiction Is Hijacking Our Kids - and How to Break the Trance* St. Martin's Griffin
- Adair, Cam (2017) *Video Game Addiction 101 for Parents & Therapists* https://www.youtube.com/watch?v=3Edhni1X5O8
- Adair, Cam *Game Quitters* https://gamequitters.com/

Chapter Four -Why Does Sleep Become a Parenting Struggle Again?

Hale, L, Kirschen, G.W., LeBourgeois, M.K., Gradisar, M, Garrison, M.M., Montgomery-Downs, H, Kirschen, H., McHale, S.M., Chang, A., Buxton, O.M. (2018) *Youth screen media habits and sleep: sleep-friendly screen-behavior recommendations for clinicians, educators, and parents* National Institute of Health https://pmc.ncbi.nlm.nih.gov/articles/PMC5839336/

Coughlan, S (2020) *Most Children Sleep with Mobile Phone Beside Bed* BBC https://www.bbc.com/news/education-512961 97

Suni, E., Cotliar, D. (2024) *Interrupted Sleep: Causes & Helpful Tips* -Sleep Foundation https://www.sleepfoundation.org/slee p-deprivation/interrupted-sleep

Wilckens, K.A., Woo, S.G., Kirk, A.R., Erickson, K.I., Wheeler, M.E. (2014) *Role of sleep continuity and total sleep time in executive function across the adult lifespan* National Institute of Health https://pubmed.ncbi.nlm.nih.gov/25244484/

Walker, Matt (2019) *Sleep Is Your Superpower* -TED -Ideas Worth Spreading https://www.youtube.com/watch?v=5MuIM qhT8DM

Lally, M, Valentine-French, S (2019) *LIFESPAN DEVELOPMENT A Psychological Perspective Second Edition* -College of Lake County Illinois https://dept.clcillinois.edu/psy/lifespandevelo pment.pdf

(2008) *Teens and Sleep Paediatr Child Health. 2008 Jan; 13(1): 69–70* https://pmc.ncbi.nlm.nih.gov/articles/PMC2528821/

Teenagers and Sleep, How Much Is Enough? Hopkins Medicine https://www.hopkinsmedicine.org/health/wellness-and-prev ention/teenagers-and-sleep-how-much-sleep-is-enough

(2024) *About Sleep* CDC https://www.cdc.gov/sleep/about/?C DC_AAref_Val=https://www.cdc.gov/sleep/about_sleep/how _much_sleep.html

(2020) *Sleep In Middle and High School Students* CDC https://archive.cdc.gov/#/details?url=https://www.cdc.gov/healthyschools/features/students-sleep.htm

Richter, Ruthann (2015) *Among teens, sleep deprivation an epidemic* Stanford Medicine https://med.stanford.edu/news/all-news/2015/10/among-teens-sleep-deprivation-an-epidemic.html

(2013) *How Sleep Clears the Brain* National Institute of Health https://www.nih.gov/news-events/nih-research-matters/how-sleep-clears-brain

Davidson, J.R. Molodfsky, H., Lue, F.A. (1991) *Growth hormone and cortisol secretion in relation to sleep and wakefulness* National Institute of Health https://pmc.ncbi.nlm.nih.gov/articles/PMC1188300/

(2013) *The Benefits of Slumber Why You Need a Good Night's Sleep* National Institute of Health https://newsinhealth.nih.gov/2013/04/benefits-slumber

(2025) *Sleep tips: 6 steps to better sleep* Mayo Clinic https://www.mayoclinic.org/healthy-lifestyle/adult-health/in-depth/sleep/art-20048379

Chapter Five -Who Is at the Other End of Technology?

United States Attorney's Office, Northern District of New York (2020) *Internet Predators: Warnings & Prevention for Families During the Pandemic and Beyond* Press Release https://www.justice.gov/usao-ndny/pr/internet-predators-warnings-prevention-families-during-pandemic-and-beyond

Watson, Graham (2013) *Manti Te'o's uncle says money and notoriety fueled the hoax* Yahoo Sports https://tinyurl.com/3n3bdnab

KOAA (2021) *FBI estimates 500,000 online predators are a daily*

threat to kids going online NBC 5 News and Yahoo News https://w
ww.yahoo.com/news/fbi-estimates-500-000-online-125145
875.html

Miller, Crista (2017) *Dangers of Technology: Indiana Mom
Shares Story of Daughter Who Became Victim to Human Trafficking
-*WNDU https://tinyurl.com/ykeecwaz

Stamm, Dan (2017) *High School Student Faces Child Porn
Charges for Extorting Nude Photos While Posing as Girl: DA* NBC
10 Philadelphia https://www.nbcphiladelphia.com/news/local/
catfish-teen-extort-nude-images/15879/

(2025) How to Protect Your Teen from Online Predators
without Controlling Their Every Move Parenting Teens with
Dr. Cam https://open.spotify.com/episode/5VpHeMY5rV2PV
YSZ8NCRRS

Pandith, F, Ware, J (2021) *Teen terrorism inspired by social
media is on the rise. Here's what we need to do.* NBC News
https://www.nbcnews.com/think/opinion/teen-terrorism-ins
pired-social-media-rise-here-s-what-we-ncna1261307

Christakis, Nicholas, and Fowler, James (2011) *Connected: The
Surprising Power of Our Social Networks and How They Shape Our
Lives — How Your Friends' Friends' Friends Affect Everything You
Feel, Think, and Do* Little Brown Spark

Sforza, Lauren (2024) *Most of Gen Z using TikTok for health
advice: Survey* The Hill https://thehill.com/policy/technology/
4774795-tiktok-gen-z-health-advice/

(2024) *Healthy or Not? Attitudes to Wellness on TikTok* Zing
Coach https://www.zing.coach/fitness-library/healthy-or-no
t-attitudes-to-wellness-on-tiktok

Herrick, Devon (2024) TikTok Therapists Promote Family
Estrangement to Young People Goodman Health
https://www.goodmanhealthblog.org/tiktok-therapists-pr

omoting-family-estrangement-therapy-to-young-people/

Barry, Ellen (2024) Is Cutting Off Your Family Good Therapy? The New York Times

https://www.nytimes.com/2024/07/14/health/therapy-fam ily-estrangement.html

Chapter Six -Ideas for Preventing Predator Connections

Haidt, Jonathan (2024) *The Anxious Generation* Penguin Press

Persin, Coby (2015) *The Dangers of Social Media (Child Predator Experiment)* https://www.youtube.com/watch?v=6jMhMVEjEQ g

(2025) How to Protect Your Teen from Online Predators without Controlling Their Every Move Parenting Teens with Dr. Cam https://open.spotify.com/episode/5VpHeMY5rV2PV YSZ8NCRRS

Clair, D, McMahon, J (1979-1988) The Facts of Life Sony Pictures Televison

Borba, Michelle (2017) Unselfie Touchstone https://www. amazon.com/UnSelfie-Empathetic-Succeed-All-About-Me-World/dp/1501110071/ref=sr_1_1?crid=13F3X0HK6YE6B&ke ywords=unselfie+book&qid=1674406033&sprefix=unsel%2C aps%2C274&sr=8-1

Chapter Seven -Why Do Kids Struggle with the Permanence of the Internet?

Atomic Entertainment (2016) *The Internet Ruined My Life* Left Right Productions https://tinyurl.com/45bsafpj

Vedantum, Shankar (2022) *You Can't Hit Unsend* Hidden Brain Podcast https://hiddenbrain.org/podcast/you-cant-hit-unse

nd/

What to Do When Your Kid Posts Inappropriate Content Online Digital Citizen's Academy https://tinyurl.com/5aa9yuyr

Worall, William (2020) *What to Do If Someone Screenshots Your Snapchat Photo* Hacked https://hacked.com/what-to-do-if-so meone-screenshots-your-snapchat-photo/

Konnikova, Maria (2014) *The Lost Art of the Unsent Angry Letter* New York Times https://www.nytimes.com/2014/03/23/opinio n/sunday/the-lost-art-of-the-unsent-angry-letter.html

Chapter Eight - The Porn Chats

Damour, Lisa (2023) *The Emotional Lives of Teenagers: Raising Connected, Capable, and Compassionate Adolescents* Ballantine Books

Paslakis, G, Chiclana Actis, C, Mestre-Bach, G (2020) *Associations between pornography exposure, body image and sexual body image: A systematic review* Journal of Health Psychology Sage Journals https://journals.sagepub.com/doi/10.1177/135910532 0967085#:~:text=Compelling%20evidence%20shows%20tha t%20frequency,women%20appear%20to%20be%20affected.

Park, B.Y. Wilson, G., Berger, J., Christman, M., Reina, B., Bishop, F., Klam, W., Doan, A.P., (2016) *Is Internet Pornography Causing Sexual Dysfunctions? A Review with Clinical Reports -* National Institute of Health https://pmc.ncbi.nlm.nih.gov/arti cles/PMC5039517/

Luscombe, B., (2016)*Porn and the Threat to Virility* Time Magazine https://time.com/magazine/us/4277492/april-11 th-2016-vol-187-no-13-u-s/

Bahrampor, T. (2016) *There Isn't Really Anything Magical About It; Why More Millenials Are Putting Off Sex* -Washington Post https://www.washingtonpost.com/local/social-issues/th

ere-isnt-really-anything-magical-about-it-why-more-mill
ennials-are-putting-off-sex/2016/08/02/e7b73d6e-37f4-11e
6-8f7c-d4c723a2becb_story.html

(2019) *Watching Pornography Rewires the Brain to a More Juvenile State* -The Conversation https://theconversation.com/watching-pornography-rewires-the-brain-to-a-more-juve nile-state-127306

Twenge, Jean (2018) *iGen: Why Today's Super-Connected Kids Are Growing Up Less Rebellious, More Tolerant, Less Happy—and Completely Unprepared for Adulthood—and What That Means for the Rest of Us* Atria

Enough is Enough https://enough.org/

Simeone, M. (2021) *Howard Stern Closes Out 2021 with Billie Eilish, Ben Affleck, Neil Young, and More* The Howard Stern Show https://enough.org/

Cybersavvykids.org

Turning Teen https://turningteen.com/

Start here: Evolution has not prepared your brain for today's porn https://www.yourbrainonporn.com/miscellaneous-resou rces/start-here-evolution-has-not-prepared-your-brain-fo r-todays-porn/

(2021) *Billie Eilish on How Watching Porn When She Was 11 DESTROYED Her Brain* Entertainment Tonight https://www.yo utube.com/watch?v=IoziCKinuQI

(2021) Billie Eilish Calls Porn a Disgrace and Destroys It - Yazzfetto https://www.youtube.com/watch?v=-PBkFHLAr38

Thorne, Jack, Graham, Stephen (2025) *Adolescence* Netflix https://www.netflix.com/title/81756069

Gillett, Francesca (2024) *Influencers driving extreme misogyny, say police* BBC https://www.bbc.com/news/articles/cne4vw1x8 3po

Jones, CT (2025) *Why 'Adolescence' Is Sparking Conversations About Incel Dread Online* Rolling Stone https://www.rollingston e.com/culture/culture-features/adolescence-netflix-manosp here-radicalization-1235301459/

Chapter Nine: The Rise of Online Gambling

Hollenbeck, Brett (2024) *The Financial Consequences of Legalized Sports Gambling | Brett Hollenbeck* https://bretthollenbeck. com/wp-content/uploads/2024/07/hollenbeck_sports_gamb ling.pdf

Ponciano, Jonathan (2025) Are Americans Betting Their Future on Sports? Uncover the Surprising Stats Investopedia https://www.investopedia.com/americans-sports-betting-l osing-8768618

Simon, Clea (2025) *Gambling problems are mushrooming. Panel says we need to act now* The Harvard Gazette https://new s.harvard.edu/gazette/story/2025/01/online-gambling-is-on-the-rise-panel-says-we-need-to-act-now/

(2024) *Teen Gambling Raises Concerns* Cbs Mornings Plus https://www.cbsnews.com/video/the-rise-of-legal-gambl ing-and-its-hidden-impact-on-teenager/

Tran, L, Wardle, H PhD, Colledge-Frisby, S, PhD, Taylor, S, MPH, Lynch, M MGH, Rehm, J, PhD (2024) *The prevalence of gambling and problematic gambling: a systematic review and meta-analysis* The Lancet https://www.thelancet.com/journal s/lanpub/article/PIIS2468-2667(24)00126-9/fulltext

Hollenbeck, B, Larsen, P, Proserpio, D (2024) *The Financial Consequences of Legalized Sports Gambling* Bretthollenbeck.com https://bretthollenbeck.com/wp-content/uploads/2024/07/ hollenbeck_sports_gambling.pdf

(2025) *Personal stories* Gamblershelp.com/au https://gamble

rshelp.com.au/learn-about-gambling/personal-stories/

Spohr, Mike (2024) *People With Gambling Addictions Hold Nothing Back About How It Unraveled Their Lives* Buzzfeed https://www.buzzfeed.com/mikespohr/people-with-gambling-addictions-hold-nothing-back

Lewis, Michael (2024-2025) *Against the Rules* Pushkin https://www.pushkin.fm/podcasts/against-the-rules

Regulatory Capture Wikipedia https://en.wikipedia.org/wiki/Regulatory_capture

Chapter Ten: Social Media and Anxiety

Orsorio, A. (2022) *Research Update: Children's Anxiety and Depression on the Rise* Georgetown University McCourt School of Public Policy Center for Children and Families https://ccf.georgetown.edu/2022/03/24/research-update-childrens-anxiety-and-depression-on-the-rise/

(2023) *The National Survey of Children's Health* -Data Resource Center for Child and Adolescent Health https://www.childhealthdata.org/learn-about-the-nsch/NSCH

Lebrun-Harris, L, Ghandour, R.M., Kogan, M (2022) *Five-Year Trends in US Children's Health and Well-being, 2016-2020* JAMA Pediatrics https://tinyurl.com/szv48ede

McCarthy, Claire () *Anxiety in Teens Is Rising, What's Going On?* Healthy Children https://www.healthychildren.org/English/health-issues/conditions/emotional-problems/Pages/Anxiety-Disorders.aspx

Thompson, Derek (2022) *Why Are American Teenagers So Sad and Anxious* -Plain English with Derek Thompson Podcast/ The Ringer https://www.theringer.com/2022/04/22/health/why-are-american-teenagers-so-sad-and-anxious

Twenge, Jean (2018) *iGen: Why Today's Super-Connected Kids*

Are Growing Up Less Rebellious, More Tolerant, Less Happy—and Completely Unprepared for Adulthood—and What That Means for the Rest of Us Atria

Haidt, Jonathan (2021) *Facebook's Dangerous Experiment on Teen Girls* The Atlantic https://www.theatlantic.com/ideas/archive/2021/11/facebooks-dangerous-experiment-teen-girls/620767/

Abrams, Zara (2023) *Why Young Brains Are Especially Vulnerable to Social Media* -American Psychological Association https://www.apa.org/news/apa/2022/social-media-children-teens

Katz, Leslie (2020) *How Tech and Social Media Are Making Us Feel Lonelier Than Ever* -CNET https://www.cnet.com/culture/features/how-tech-and-social-media-are-making-us-feel-lonelier-than-ever/

(2022) *How to Cope with News Anxiety* -Mental Health UK https://mentalhealth-https://tinyurl.com/bdeunm2f

Levitin, Daniel (2015) *The Organized Mind* Dutton

Levitin, Daniel (2015) *Daniel Levitin on Information Overload* -RSA https://www.youtube.com/watch?v=LjeKUsBe5UU&t=15s

Thompson, Derek (2024) *A Psychologist Explains Four Reasons the Internet Feels So Broken* -Plain English with Derek Thompson Podcast on the Ringer https://www.theringer.com/2024/04/09/pop-culture/psychologist-jay-van-bavel-explains-four-reasons-the-internet-feels-broken-negativity-bias

Schuessler, Jennifer (2024) *Brain Rot* -New York Times https://www.nytimes.com/2024/12/01/arts/brain-rot-oxford-word.html

Chapter Eleven: Information Overload •Attention Span

U2 (1991) *Until the End of the World* Island Records

Pakman, David (2023) *How Lack of Attention Span Increases Anxiety (Dr. Gloria Mark Interview)* -David Pakman Show https://www.youtube.com/watch?v=ONc9vSgoTCc

Yeykelis, L., Cummings, J., Reeves, B. (2014) *Multitasking on a Single Device: Arousal and the Frequency, Anticipation, and Prediction of Switching Between Media Content on a Computer* Journal of Communication https://tinyurl.com/358bbj94

Levitin, Daniel (2015) *The Organized Mind* Dutton

Haidt, Jonathan (2022) *Why the Past Ten Years of American Life Have Been Uniquely Stupid* The Atlantic https://www.theatl antic.com/magazine/archive/2022/05/social-media-democra cy-trust-babel/629369/

Davis, Jennifer E (2023) *Multitasking and How It Affects Your Brain Health* Lifespan https://tinyurl.com/zyydpd5m

Stillman, Jessica (2020) *New Study: Multitasking Is Making Your Anxiety WorseThe more you switch tasks, the sadder and more fearful you feel, new research shows* -Inc

Capuano, Cara (2023) Can't pay attention? You're not alone University of California Irvine https://www.universityofcalifornia.edu/news/cant-pay-att ention-youre-not-alone

Wolf, Maryanne (2018) Skim reading is the new normal. The effect on society is profound The Guardian https://www.thegu ardian.com/commentisfree/2018/aug/25/skim-reading-new-normal-maryanne-wolf

Quintero, E, and Brennan-Gac, T (2024) The Threat of Technology to Students' Reading Brains The Shanker Institute https://www.shankerinstitute.org/blog/technology-and-re ading#:~:text=Research%20suggests%20that%20the%20ha bits,notifications%20and%20algorithm%2Dgenerated%20su ggestions

Ansari, Nyel (2024) The Impact Of Short Form Content TEDxTalks
    https://www.youtube.com/watch?app=desktop&v=hPRDA M3BCK0

Goodpaster, Caitlyn (2023) How TikTok Hijacks Your Brain Knowing Neurons
    https://knowingneurons.com/blog/2023/06/20/how-tiktok -hijacks-your-brain/

Kumar, Karthik How Long Should I Meditate to See Results? MedicineNet
    https://www.medicinenet.com/how_long_should_i_medi tate_to_see_results/article.htm

Chapter Twelve - The Time Displacement of Well-Being Activities

Richter, Ruthann (2015) *Among teens, sleep deprivation an epidemic* Stanford Medicine https://med.stanford.edu/news/al l-news/2015/10/among-teens-sleep-deprivation-an-epidem ic.html

Newport, Cal (2019) *Digital Minimalism, Choosing Focus in a Noisy World* Portfolio

Twenge, Jean (2018) *iGen: Why Today's Super-Connected Kids Are Growing Up Less Rebellious, More Tolerant, Less Happy—and Completely Unprepared for Adulthood—and What That Means for the Rest of Us* Atria

Sellman, Mark (2025) *Students will spend 25 years of their lives on their mobiles* The Times https://www.thetimes.com/ uk/technology-uk/article/average-young-person-25-years- phone-screen-time-hwt76mnpq#:~:text=Daily%20average %20screen%20time%20increased,per%20cent%20among%

20university%20students

Rubin, Gretchen (2024) *Life in Five Senses: How Exploring the Senses Got Me Out of My Head and Into the World* Crown Publishing Group

Cline, Ernest (2012) *Ready Player One: A Novel* Ballantine Books

Csikszentmihalyi, Mihaly (2004) *Flow, the secret to happiness* Ted https://www.ted.com/talks/mihaly_csikszentmihalyi_flo w_the_secret_to_happiness

Csikszentmihalyi, Mihaly (1990) *FLOW: The Psychology of Optimal Experience* Harper and Row /Global Learning Communities /The Way Back Machine https://web.archive.org/web /20150225201519/http://www.psy-flow.com/sites/psy-flow/fi les/docs/flow.pdf

Kardaras, Nicholas (2017) *Glow Kids: How Screen Addiction Is Hijacking Our Kids - and How to Break the Trance* St. Martin's Griffin

Chapter Thirteen - Individuation and Lack of Autonomy

Haidt, Jonathan (2022) *Why the Past Ten Years of American Life Have Been Uniquely Stupid* The Atlantic https://www.theatl antic.com/magazine/archive/2022/05/social-media-democra cy-trust-babel/629369/

Gray, Peter (2023) The Special Value of Age-Mixed Play I: How Age Mixing Promotes Learning
https://petergray.substack.com/p/10-the-special-value-of-age-mixed

Hu, Charlotte (2024) Why Writing by Hand Is Better for Memory and Learning Scientific American
https://www.scientificamerican.com/article/why-writing-b

y-hand-is-better-for-memory-and-learning/#:~:text=A%2
0recent%20study%20in%20Frontiers,vision%2C%20sensory
%20processing%20and%20memory

Saikia, A., Das, J., Barman, P., Bharali, M., (2019) *Internet Addiction and its Relationships with Depression, Anxiety, and Stress in Urban Adolescents of Kamrup District, Assam* -National Institute of Health https://pmc.ncbi.nlm.nih.gov/articles/PMC 6515762/

DeFrank, Molly (2022) *Digital Detox, the Two-Week Tech Reset for Kids* Bethany House Publishers

Dunckly, Victoria (2015) *Reset Your Child's Brain: A Four-Week Plan to End Meltdowns, Raise Grades, and Boost Social Skills by Reversing the Effects of Electronic Screen-Time* New World Library

Adair, Cam (2018) *Video Game Addiction 101 for Parents & Therapists* Game Quitters https://www.youtube.com/watch? v=3Edhni1X5O8

Chapter Fourteen: Empathy and Distress Tolerance

How Much of Communication Is Nonverbal? -University of Texas Permian Basin https://online.utpb.edu/about-us/article s/communication/how-much-of-communication-is-nonver bal/

Turkle, Sherry (2016) *Reclaiming Conversation: The Power of Talk in a Digital Age* Penguin Books

Ellwood, Beth (2022) *People who frequently play violent video games like Call of Duty show neural desensitization to painful images, according to study* PsyPost https://www.psypost.org /people-who-frequently-play-call-of-duty-show-neural-de sensitization-to-painful-images-according-to-study/

Ruston, Delaney (2020) *Parenting in the Screen Age.* Starhouse Media

Coltrera, Francesca (2018) *Anxiety in children* Harvard Health https://www.health.harvard.edu/blog/anxiety-in-children-2018081414532

ADL (2023) <u>*How Tweens Experience Cyberbullying*</u> https://www.adl.org/resources/tools-and-strategies/how-tweens-experience-cyberbullying

Parrish, Shane, Huberman, Andrew (2022) *How to Control Your Impulses So You Don't Ruin Your Life* https://www.youtube.com/watch?v=8wpP1W8eoaI

Newport, Cal (2019) *Digital Minimalism, Choosing Focus in a Noisy World* Portfolio

Twenge, Jean (2018) *iGen: Why Today's Super-Connected Kids Are Growing Up Less Rebellious, More Tolerant, Less Happy—and Completely Unprepared for Adulthood—and What That Means for the Rest of Us* Atria Books

Chapter Fifteen: Connection Disruptor, Social Contagion

Redisch, Heather Adulting 101 Masterclass https://www.adulting101masterclass.com/a101-training

Wagner, D, Cin, S, Sargent, J, Kelly, W, Heatherton, T (2011) *Spontaneous Action Representation in Smokers when Watching Movie Characters Smoke* The Journal of Neuroscience

MacGeown, L, Davis, R (2018) *Social modeling of eating mediated by mirror neuron activity: A causal model moderated by frontal asymmetry and BMI* PubMed

Christakis, Nicholas, and Fowler, James (2011) *Connected: The Surprising Power of Our Social Networks and How They Shape Our Lives — How Your Friends' Friends' Friends Affect Everything You Feel, Think, and Do* Little Brown Spark

Perry, Bruce, and Winfrey, Oprah (2021) What Happened to You?: Conversations on Trauma, Resilience, and Healing Flatiron Books

https://www.amazon.com/What-Happened-You-Understanding-Resilience/dp/1250223180

Ayal, Nir (2019) *Indistractable: How to Control Your Attention and Choose Your Life* BenBella Books

Chapter Sixteen Parent's Guide to Kids' Social Media

Cooper, Anderson *Brain Hacking* -60 Minutes https://www.youtube.com/watch?v=awAMTQZmvPE

Orlowski, Jeff, Coombe, Davis, Curtis, Vickie, (2020) *The Social Dilemma* Netflix Docudrama Featuring Tristan Harris https://www.netflix.com/title/81254224?source=35

Green, Jonathan Ignatius (2018) *Social Animals* Kanopy Documentary https://www.kanopy.com/en/product/social-animals?frontend=kui

(2022) *Dirty Dozen List* National Center on Sexual Exploitation https://endsexualexploitation.org/dirtydozenlist-2022/#ddl-list

(2024) *Legislation to Protect Kids Online Passes Senate by Overwhelming Majority* CBS News https://www.cbsnews.com/news/kids-online-safety-bill-passes-senate-blumenthal-blackburn/

Koetsier, John (2024) GenZ Dumping Google For TikTok, Instagram As Social Search Wins Forbes

https://www.forbes.com/sites/johnkoetsier/2024/03/11/genz-dumping-google-for-tiktok-instagram-as-social-search-wins/

George, Anita (2022) As it turns out, TikTok isn't a reliable search engine for news Digital Trends https://www.digitaltrends.com/social-media/tiktok-isnt-a-reliable-search-engine-for-news/

Huang, Kalley (2022) For Gen Z, TikTok Is the New Search Engine The New York Times
https://www.nytimes.com/2022/09/16/technology/gen-z-tiktok-search-engine.html

Abrams, Zara (2021) *How Can We Minimize Instagram's Harmful Effects?* -American Psychological Association https://tinyurl.com/5cu4kbs3

Isaac, Mike, Singer, Natasha (2024) *Instagram, Facing Pressure Over Child Safety Online, Unveils Sweeping Changes* The New York Times https://www.nytimes.com/2024/09/17/technology/instagram-teens-safety-privacy-changes.html

Ryles, Gemma (2021) *Is TikTok Safe?* -Trusted Reviews https://www.trustedreviews.com/news/is-tiktok-safe-417 2063

(2020) *TikTok Failed to Ban Flagged Child Predator* BBC News

Baucom, Jackie (2023, 2025) *Dangerous and Deadly TikTok Trends* -Gabb https://gabb.com/blog/tiktok-trends/?srsltid=A fmBOorV2JS-QG422VdFcEy9RqZs28P58T0X2SDiisJ9qPM39aI XnFQi

Barry, Ellen (2024) *Is Cutting Off Your Family Good Therapy?* The New York Times https://www.nytimes.com/2024/07/14/h ealth/therapy-family-estrangement.html

(2021) *Investigation: How TikTok's Algorithm Figures Out Your Deepest Desires* -Wall Street Journal https://www.wsj.com/vide o/series/inside-tiktoks-highly-secretive-algorithm/investiga tion-how-tiktok-algorithm-figures-out-your-deepest-desir es/6C0C2040-FF25-4827-8528-2BD6612E3796

Kelvin, Hadero, Haleluya (2023) *Why Tik Tok's Security Risk Keeps Raising Fears* -AP News https://apnews.com/article/tikt

ok-ceo-shou-zi-chew-security-risk-cc36f36801d84fc065211
2fa461ef140

Browning, Kellen (2021) *How Discord, Born From an Obscure Game, Became a Social Hub for Young People* New York Times https://www.nytimes.com/2021/12/29/business/discord-serv er-social-media.html?action=click&module=RelatedLinks&p gtype=Article

Zapal, Haley (2022) *The 12 Most Dangerous Apps for Kids* Bark https://www.bark.us/blog/dangerous-apps-kids/

Kelly, Samantha Murphy (2022) *The Dark Side of Discord for Teens* CNN Business https://www.cnn.com/2022/03/22/tech/d iscord-teens/index.html

Guyer, Jonathan (2023) *The Ongoing Scandal Over Leaked US Intel Documents, Explained* Vox https://www.vox.com/world-p olitics/2023/4/10/23677820/leaked-intelligence-documents-ukraine-war-discord-4chan

Porter, Matt (2019) *Mr Beast explains why he won't replace young streamer's destroyed PC* Dextero https://www.dexerto.co m/entertainment/mr-beast-explains-why-he-wont-replace-young-streamers-destroyed-pc-322804/

Knorr, Caroline (2021) *Parents' Ultimate Guide to YouTube Kids* Common Sense Media https://www.commonsensemedia.org/ articles/parents-ultimate-guide-to-youtube-kids?check_log ged_in=1

Winck, Ben (2021) *How Reddit Day Traders Are Using the Platform to Upend the Stock Market and Make Money In the Process* Business Insider https://markets.businessinsider.com/news/s tocks/reddit-day-traders-wallstreetbets-gamestop-gme-rall y-upending-stock-market-2021-1-1030009280

(2019) *Is Reddit the Most Influential Site on the Internet?* Knowledge at Wharton Interview with Christine Lagorio-

Chafkin

Troutner, Allison (2024) *Is Reddit Safe? Tips for Parents to Keep Teens Safer Online* -VPN Overview https://vpnoverview.com/internet-safety/kids-online/is-reddit-safe/

Chapter Seventeen -Ideas for Social Media Use

ToSDR *Terms of Service Didn't Read* https://tosdr.org/en

Abrams, Zara (2021) *How Can We Minimize Instagram's Harmful Effects?* -American Psychological Association https://tinyurl.com/5cu4kbs3

Alter, Adam (2018) *Irresistible, the Rise of Screens and the War for Attention* Penguin Books

Johansen, Alison Grace (2023) *Should You Accept Cookies? 5 Times You Definitely Shouldn't* Norton

Chapter Eighteen -The AI Conversation

De Kai, (2025) *Raising AI: An Essential Guide to Parenting Our Future* The MIT Press https://www.amazon.com/s?k=Raising+AI&crid=3TD78YHR4SUUN&sprefix=raising+ai+%2Caps%2C179&ref=nb_sb_noss_2

(2019) *Fake news can give us false memories* University College Cork https://www.ucc.ie/en/news/2019/fake-news-can-give-us-false-memories-study-finds.html

Wagner, U, Schlechter, P, Echterhoff, G (2019) *Socially Induced False Memories in the Absence of Misinformation* Nature.com https://www.nature.com/articles/s41598-022-11749-w

Explained: What Are Deepfakes? -Webwise.ie https://tinyurl.com/mr32njph

Phan, Karena (2019) Fakeout -Time for Kids https://www.timeforkids.com/g56/fakeout-2/

Landau, Shira (2021) Teaching Kids about Deep Fake Technologies Elearning Industry https://elearningindustry.com/teaching-children-about-deepfake-technologies

A Beginners Guide to Deepfakes Our Safer Schools UK https://oursaferschools.co.uk/2022/07/20/a-beginners-guide-to-deepfakes/

(2024) *Parents sue over AI chatbot they say encourages children to harm themselves and others* NBC Boston https://www.youtube.com/watch?v=tl_HmGkMQxw

Jasnow, Dan (2025) *New Lawsuits Targeting Personalized AI Chatbots Highlight Need for AI Quality Assurance and Safety Standards* -The National Law Review https://natlawreview.com/article/new-lawsuits-targeting-personalized-ai-chatbots-highlight-need-ai-quality-assurance

Allyn, Bobby (2024) *Lawsuit: A chatbot hinted a kid should kill his parents over screen time limits* -NPR https://www.npr.org/2024/12/10/nx-s1-5222574/kids-character-ai-lawsuit

Cooper, Anderson (2024) *Schools face a new threat: "nudify" sites that use AI to create realistic, revealing images of classmates* -60 Minutes https://www.cbsnews.com/news/schools-face-new-threat-nudify-sites-use-ai-create-realistic-revealing-images-60-minutes-transcript/

Harris, Tristan, and Stein, Zak (2026) *Attachment Hacking and the Rise of AI Psychosis* Center for Humane Technology https://centerforhumanetechnology.substack.com/p/attachment-hacking-and-the-rise-of?utm_source=substack&utm_medium=email

Fegan, Sasha (2026) Welcome to the Attachment Economy Center for Humane Technology https://centerforhumanetechnology.substack.com/p/welcome-to-the-attachment-economy

Zao-Sanders, Marc (2025) How People Are Really Using Gen AI in 2025 Harvard Business Review https://hbr.org/2025/04/how-people-are-really-using-gen-ai-in-2025

Resources for conversation on Deepfakes: Resources for Conversation:

Webwise.ie: Explained: What Are Deepfakes?
Time for Kids: Fakeout
Elearning Industry: Teaching Kids about Deep Fake Technologies
Our Safer Schools UK: A Beginners Guide to Deepfakes

Technology Website Resources
Wait Until 8th https://www.waituntil8th.org/resources
Enough Is Enough https://enough.org/
Protect Young Eyes https://www.protectyoungeyes.com/
Child Health Data https://www.childhealthdata.org/learn-about-the-nsch/NSCH
Cybersavvykids.org https://savvycyberkids.org/
yourbrainonporn.org https://www.yourbrainonporn.com/miscellaneous-resources/start-here-evolution-has-not-prepared-your-brain-for-todays-porn/
Center for Humane Technology https://www.humanetech.com/families-educators
Dr Kristy Goodwin https://drkristygoodwin.com/
Game Quitters https://gamequitters.com/

National Center on Sexual Exploitation https://endsexuale xploitation.org/

Internet Matters.org https://www.internetmatters.org/

Rescue Time https://www.rescuetime.com/

Turning Teen https://turningteen.com/

Take it Down https://takeitdown.ncmec.org/

ToSDR *Terms of Service Didn't Read* https://tosdr.org/en

Digital Citizen Academy https://digitalcitizenacademy.org/

Adulting 101 Masterclass https://www.adulting101mastercla ss.com/a101-training

Technology Book Resources:

Turkle, Sherry (2016) *Reclaiming Conversation: The Power of Talk in a Digital Age* Penguin Books

Newport, Cal (2019) *Digital Minimalism, Choosing Focus in a Noisy World* Portfolio

DeFrank, Molly (2022) *Digital Detox, the Two-Week Tech Reset for Kids* Bethany House Publishers

Dunckly, Victoria (2015) *Reset Your Child's Brain: A Four-Week Plan to End Meltdowns, Raise Grades, and Boost Social Skills by Reversing the Effects of Electronic Screen-Time* New World Library

Ruston, Delaney (2020) *Parenting in the Screen Age.* Starhouse Media

Jenson, Kristen A (2014) *Good Pictures Bad Pictures: Porn-Proofing Today's Young Kids*

Weeks, Jennifer (2016) *The New Age of Sex Education:: How to Talk to Your Teen About Cybersex and Pornography in the Digital Age (1)* Book Baby

Alter, Adam (2018) *Irresistible, the Rise of Screens and the War for Attention* Penguin Books

Levitin, Daniel (2015) *The Organized Mind* Dutton

Lagorio-Chafkin, Christine (2018) *We Are the Nerds: The Birth and Tumultuous Life of Reddit, the Internet's Culture Laboratory* Grand Central Publishing

Twenge, Jean (2018) *iGen: Why Today's Super-Connected Kids Are Growing Up Less Rebellious, More Tolerant, Less Happy—and Completely Unprepared for Adulthood—and What That Means for the Rest of Us* Atria Books

Sales, Nancy Jo (2017) *American Girls: Social Media and the Secret Lives of Teenagers* Vintage

Haidt, Jonathan (2024) *The Anxious Generation* Penguin Press

Damour, Lisa (2023) *The Emotional Lives of Teenagers: Raising Connected, Capable, and Compassionate Adolescents* Ballantine Books

Ayal, Nir (2019) *Indistractable: How to Control Your Attention and Choose Your Life* BenBella Books

Kardaras, Nicholas (2017) *Glow Kids: How Screen Addiction Is Hijacking Our Kids - and How to Break the Trance* St. Martin's Griffin

Technology Movie/Show Resources

Ruston, Delaney (2016) *Screenagers* (Screenagers)

Ruston, Delaney (2019) *Screenagers Next Chapter* Screenagers Next Chapter

Orlowski, Jeff, Coombe, Davis, Curtis, Vickie, (2020) *The Social Dilemma* Netflix Docudrama Featuring Tristan Harris https://www.netflix.com/title/81254224?source=35

Green, Jonathan Ignatius (2018) *Social Animals* Kanopy Documentary https://www.kanopy.com/en/product/social-animals?frontend=kui

Schulman, Nev (2013) *Catfishing* https://tinyurl.com/3yjapst t

Atomic Entertainment (2016) *The Internet Ruined My Life* Left Right Productions https://tinyurl.com/45bsafpj

Technology Video/ Podcast Resources:

Alter, Adam (2020) *Think Better* -University of California Booth School of Business https://www.youtube.com/watch?v= -RmoWCHZjt4

Adair, Cam (2018) *Video Game Addiction 101 for Parents & Therapists* Game Quitters https://www.youtube.com/watch?v=3Edhni1X5O8

Cooper, Anderson *Brain Hacking* -60 Minutes https://www.youtube.com/watch?v=awAMTQZmvPE

Walker, Matt (2019) *Sleep Is Your Superpower* -TED -Ideas Worth Spreading https://www.youtube.com/watch?v=5MuIMqhT8DM

Thompson, Derek (2022) *Why Are American Teenagers So Sad and Anxious* -Plain English with Derek Thompson Podcast on the Ringer

Levitin, Daniel (2015) *Daniel Levitin on Information Overload* -RSA

Persin, Coby (2015) *The Dangers of Social Media (Child Predator Experiment)* https://www.youtube.com/watch?v=6jMhMVEjEQg

Vedantum, Shankar (2022) *You Can't Hit Unsend* Hidden Brain Podcast https://hiddenbrain.org/podcast/you-cant-hit-unsend/

Simeone, M. (2021) *Howard Stern Closes Out 2021 with Billie Eilish, Ben Affleck, Neil Young, and More* The Howard Stern Show https://enough.org/Eilish -Sirius xm

(2021) *Billie Eilish on How Watching Porn When She Was 11 DESTROYED Her Brain* Entertainment Tonight https://www.yo utube.com/watch?v=IoziCKinuQI

(2021) *Billie Eilish Calls Porn a Disgrace and Destroys It -* Yazzfetto https://www.youtube.com/watch?v=-PBkFHLAr38

(2021) *Investigation: How TikTok's Algorithm Figures Out Your Deepest Desires* -Wall Street Journal https://www.wsj.com/vide o/series/inside-tiktoks-highly-secretive-algorithm/investiga tion-how-tiktok-algorithm-figures-out-your-deepest-desir es/6C0C2040-FF25-4827-8528-2BD6612E3796

(2019) *Is Reddit the Most Influential Site on the Internet?* Knowledge at Wharton Interview with Christine Lagorio-Chafkin

Pakman, David (2023) *How Lack of Attention Span Increases Anxiety (Dr. Gloria Mark Interview)* -David Pakman Show https://www.youtube.com/watch?v=ONc9vSgoTCc

(2024) Parents sue over AI chatbot they say encourages children to harm themselves and others -NBC Boston https://w ww.youtube.com/watch?v=tl_HmGkMQxw

Cooper, Anderson (2024) *Schools face a new threat: "nudify" sites that use AI to create realistic, revealing images of classmates* -60 Minutes

Connect

Now, more than ever, we need a supportive community to lean on while navigating parenting in the modern world. My kids would proclaim I was the **only** parent who limited screen use. I couldn't keep up with the kids' advanced knowledge of tech and ability to find workarounds to some rules.

The kids planted seeds of doubt -*was* I the only parent with qualms about big tech's impact on kids? I wished for a place where I could commiserate and collaborate with families on tech guidance and all the other hurdles the modern world throws at families. The Whyfully Community is that place.

Whyfully's Mission: Spark Connection:

- Hone a lasting connection within families -strengthen the family bond
- Actionable ideas to coach kids' (and adults) connection-building skills
- Connection to the community and world diminishes individual stress, anxiety, and self-involvement
- Connect with fellow parents on the modern parenting journey

Share Your Story: Every parent's journey is unique, and your story can inspire others. Whether it's a recent triumph or a lesson learned, your experiences are invaluable.

Learn From Others: Find comfort and guidance in the array of experiences shared by fellow community members. The diverse perspectives and ideas can provide helpful insights that may not be top-of-mind.

Access Resources: Our platform offers a wealth of resources tailored to support you in various aspects of parenting. From thoroughly researched content to practical ideas, you have a library of support at your fingertips.

Collaboration: Transform with collective wisdom and experience and enrich each others' lives with meaningful impact. Build lasting relationships with others who understand the joys and challenges of parenthood. Connect with your people who share interests and values.

We hope to connect with you there and get to know you!

Join the conversation at https://whyfully.mn.co

Book Group Discussion Ideas

1. Which concept resonated well with you?
   - Communication Ideas
   - Behavioral Psychology to draw us to our apps and devices
   - Tech's impact on sleep
   - Wariness of people we meet online
   - Pause before posting or responding
   - Modern challenges with porn or gambling through devices
   - Social Media impacts
   - Information overload and attention span
   - Impact on happiness activities
   - Effect on Empathy and Distress Tolerance
   - Connection Disruption
   - AI
2. Has tech use displaced any creative pursuits or in-person time?
3. What positives does modern tech bring to your life?
4. What do you think about tech's impacts on empathy or distress tolerance?
5. Have you felt a connection disrupted by others' device use? Do you do or say anything when this happens? Is it a small or big problem in your life?
6. Do you think modern tech challenges might impact the

economy? (i.e., gambling, devices eating into productivity time)

7. Have you ever felt like you knew someone that you don't actually know (like Nora's Molly Ringwald story)?

8. What do you think about AI and how it might impact our lives?

9. How does modern tech make your life easier? How does it make it harder?

10. Can you think of any example of growing individuation from devices?

11. Has tech ever impacted your sleep?

12. Have you ever sent a message and then wondered if you should have worded it differently? Worried about the reception of the message? Wished you could unsend it?

13. Do you feel like your head is reaching capacity for information some days? Does it feel like your memory is slipping? Your attention span shortening?

14. Are there action steps you might take to negate some of modern tech's downsides?

About the Author

This is **Nora Duncan O'Brien's** first book. Nora is the founder of Whyfully, a community where families connect to navigate modern world challenges together. After years of learning about human behavior on the front lines of the retail and sales executive worlds, she returned to her undergraduate training in research communications. Nora spent the last several years researching and writing on topics to reignite connection. Nora lives in Chicago with her husband and two boys.

Find whyfully at whyfully.com

You can connect with me on:

🌐 https://whyfully.com
📘 https://www.facebook.com/profile.php?id=100091935207222
🔗 https://www.linkedin.com/in/nora-o-brien-76a1489
🔗 https://www.instagram.com/whyfully
🔗 https://whyfully.mn.co

Subscribe to my newsletter:

✉ https://whyfully.kit.com/336ef49df4

www.ingramcontent.com/pod-product-compliance
Lightning Source LLC
Chambersburg PA
CBHW071724120626
46550CB00002B/371